AF606549

ALLE ZEIT WACH
1842

ADVANCES IN BIOCHEMICAL ENGINEERING

Volume 4

Editors: T. K. Ghose, A. Fiechter, N. Blakebrough

Managing Editor: A. Fiechter

With 87 Figures

Springer-Verlag
Berlin · Heidelberg · New York 1976

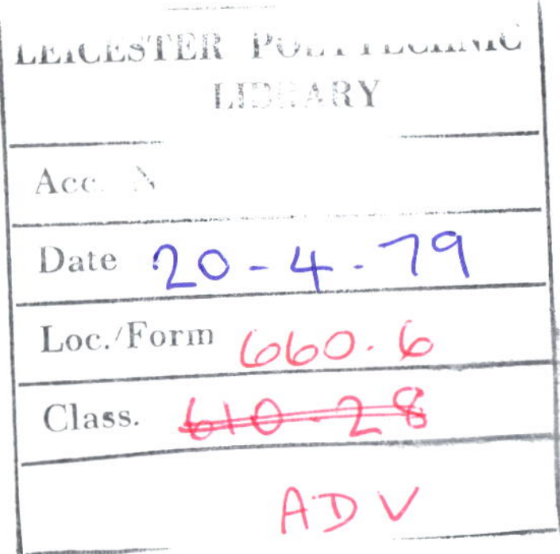

ISBN 3-540-07747-2 Springer-Verlag Berlin · Heidelberg · New York
ISBN 0-387-07747-2 Springer-Verlag New York · Heidelberg · Berlin

Library of Congress Cataloging in Publication Data: Advances in biochemical engineering. 1— . 1971— . Berlin, New York, Springer-Verlag. v. illus. 24 cm. 1. Biochemical engineering. Collected works. TP248.3.A38. 660'.2. 72-152360.

Typesetting, printing, and bookbinding: Brühlsche Universitätsdruckerei Gießen.

Contents

Editorial Guidelines

The aim of this series is to keep bioengineers and microbiologists informed of the fundaments and advances pertaining to the biochemical processes they need for the construction of bio-plants—be they for water purification, obtaining enzymes or antibiotics, for breeding yeasts, or those required for other special biochemical or biosynthetic operations. This series will likewise familiarize the biochemist with how the engineer thinks and proceeds in his work, as well as with the constructive aids at his disposal. Providing the various specialists with such extensive information is not an easy task: the backgrounds of the biochemist, the microbiologist, and the engineer are founded on entirely different bases; yet they must work side-by-side in the constantly changing field of biochemical engineering.

With this as foremost consideration, the Editors will make a special effort to present a selection of premises as well as new findings and ways of applying innovations that arise. The field of biochemical engineering is still developing and making advancements in highly industrialized nations; it is also becoming increasingly significant in those lands plagued by food shortages, which are still wrestling with problems of development today. Of primary interest for these countries are discoveries of methods for obtaining valuable natural substances and for disposing of wastes—where possible, recycling them into useful and even highly beneficial products. Advances in Biochemical Engineering can provide them with relevant contributions dealing with means of supplying food—proteins, in particular. Therefore, just as biochemistry and technology are brought together in this series, the reader will be offered contributions from industrial nations and from those countries that are presently in need of progress in the area of technology.

The Editors look forward to a strong influx of manuscripts and will do their utmost to insure the series' rapid publication. They will be published in English in order to afford the widest possbile outreach. Editors and Editorial Board are now prepared to accept manuscripts for consideration.

The Editors

CHAPTER 1

Transfer of Oxygen and Scale-Up in Submerged Aerobic Fermentation

Y. MIURA, Dept. of Biochemical Engineering, Faculty of Pharmaceutical Sciences, Osaka University, 133-1 Yamadakami, Suita, Osaka, Japan

With 20 Figures

Contents

1. Introduction

The rationale in fermenter design and operation requires a wealth of information on the factors relative to oxygen transfer. Submerged aerobic fermentations are now ubiquitous in the development of new antibiotics and in practically all other fermentations of industrial significance.

The provision of an oxygen supply adequate to meet the metabolic requirements of the organism employed is absolutely essential to the successful use of the techniques. In the evaluation of oxygen transfer in fermentation, it is necessary to evaluate the resistances to the transfer of oxygen. The overall oxygen-transfer mechanism can be divided into a series of individual rate processes as follows: diffusional transfer in gas and liquid films around gas bubbles, liquid path, liquid film around the cell, and the intracellular biochemical reaction. It was shown by Calderbank (1959b) and Yoshida *et al.* (1960) that the gas-film resistance to gas absorption in an agitated gas-liquid contactor was negligible in comparison with the liquid-film resistance, and the resistance due to eddy diffusion in the liquid path was also negligible. In the present paper, oxygen transfer and scale-up in submerged aerobic fermentation are discussed in relation to the cultivation of single-cell organisms and multicellular organisms, especially when they form mycelial pellets.

2. Analysis of Oxygen Transfer in Submerged Aerobic Fermentation

2.1 In Cultivation of Single-Cell Organisms

The rate of change of oxygen concentration in liquid medium can be expressed by the following equation:

$$\frac{dC}{d\theta} = k_L a(C_L^* - C_L) - k_r C_m. \qquad (1)$$

For gas absorption in bubbling gas-liquid contactors, especially for a sparingly soluble gas like oxygen, the gas-film resistance can be negligible, i.e., the overall resistance is in the liquid film. The oxygen concentration at the gas-liquid interface is nearly equal to its solubility. In Eq. (1), the oxygen concentration in the bulk of liquid medium, C_L, equilibrium concentration, C_L^*, and oxygen-uptake rate by the organisms, $k_r C_m$, can be determined and the rate of change in oxygen concentration, $dC_L/d\theta$, can be estimated from the observed variation of C_L vs. θ. Substituting these values into Eq. (1), the volumetric oxygen-transfer coefficient around the air bubbles, $k_L a$, can be assessed. Thus, it is possible to evaluate the liquid-phase oxygen-transfer coefficient per unit of gas/liquid interfacial area, k_L, from the value of the volumetric coefficient, $k_L a$, and a is evaluated as previously decribed (Miura, 1961; Yoshida and Miura, 1963).
At steady state, the following equation describes oxygen transfer from the liquid medium to the cell.

$$k_r C_m = k_m a_m (C_L - C_{Li}). \qquad (2)$$

The terms $k_r C_m$, C_L, and a_m in Eq. (2) can be determined experimentally, but the oxygen concentration at the liquid/cell interface, C_{Li}, is beyond the usual techniques of measurement. This implies difficulty in assessing the value of k_m through Eq. (2). Assuming that the cells are spherical and taking the material balance of oxygen on a differential volume of liquid-phase around cells at steady state, the following equation can be derived:

$$\frac{d^2C}{dr^2} + \frac{2}{r}\frac{dC}{dr} = 0. \qquad (3)$$

Boundary conditions are

$$r = r_0, \qquad D_L \frac{dC}{dr} = \frac{k_r C_m}{a_m} \tag{4}$$

$$r = r_L \qquad C = C_L$$

The solution to Eq. (3) so as to satisfy boundary conditions of Eq. (4) yields

$$C = C_L - \frac{k_r C_m r^2_0}{D_L a_m} \left(\frac{r_L - r}{r_L r}\right). \tag{5}$$

Consequently, C_{Li} is given by

$$C_{Li} = C_L - \frac{k_r C_m r_0}{D_L a_m} \left(\frac{r_L - r_0}{r_L}\right). \tag{6}$$

Substituting C_{Li}, Eq. (6), into Eq. (2), the term of k_m, as defined in Eq. (2) is rearranged as follows:

$$k_m = \frac{D_L}{r_0} \left(\frac{r_L}{r_L - r_0}\right). \tag{7}$$

Because of the lack of information on the thickness of liquid films around cells, C_{Li} and k_m cannot be assessed by Eq. (6) and (7).
In order to avoid this difficulty, the following boundary conditions may be employed:

$$r = r_0, \qquad D_L \frac{dC}{dr} = \frac{k_r C_m}{a_m}, \tag{8}$$

$$r = \infty \qquad C = C_L.$$

Substitution in Eq. (3) yields

$$C = C_L - \frac{k_r C_m r^2_0}{D_L a_m r}. \tag{9}$$

Consequently, C_{Li} and k_m can be expressed by the following equations:

$$C_{Li} = C_L - \frac{k_r C_m r_0}{D_L a_m}, \tag{10}$$

$$k_m = D_L/r_0. \tag{11}$$

Because all the terms on the right-hand sides of Eqs. (10) and (11) can be determined by experiment, C_{Li} and k_m can be evaluated through Eqs. (10) and (11). But the values of C_{Li} and k_m evaluated by Eqs. (10) and (11) are less than the actual values of C_{Li} and k_m estimated by Eqs. (6) and (7), because the value of $r_L/(r_L - r_0)$ on the right-hand sides of Eqs. (6) and (7) is greater than unity.

Satterfield (1970) demonstrated that the experimental data of Brian and Hales (1969) and Levich (1962) on mass transfer to a single solid sphere moving in a liquid at low velocity is well represented by the following equation:

$$\left(\frac{k_m d_p}{D_L}\right)^2 = 4.0 + 1.21 N_{Pe}^{2/3}. \tag{12}$$

The relative velocity between a particle and fluid in an agitated tank is roughly that of free fall due to gravity, the terminal velocity of small spheres in free fall being given by the Stokes' law:

$$U_t = \frac{g d_p^2 \Delta\rho}{18\mu_L} \tag{13}$$

The corresponding Peclet number, denoted by N_{Pe}^*, is expressed by the following equation:

$$N_{Pe}^* = \frac{g d_p^3 \Delta\rho}{18\mu_L D_L}. \tag{14}$$

Bacteria and yeasts are small and the corresponding Peclet numbers obtained by Eq. (14) are so small that the second term in the right-hand side of Eq. (12) is negligible. Accordingly, the value of k_m obtained from Eq. (12) is nearly equal to that from Eq. (11). In an agitated, aerated fermenter, liquid turbulence should promote mass transfer, and the actual value of k_m is expected to be greater than that calculated from Eq. (12). According to Harriott (1962), the ratio of the k_m obtained to those estimated from Eq. (12) was between 1 and 4.

The rate of oxygen uptake due to respiration is not constant throughout the fermentation cycle but exhibits appreciable variations. It is assumed that the respiration rate is independent of oxygen concentration down to a very low value of C_L, namely the critical oxygen concentration. The respiratory oxygen-transfer coefficient per unit of liquid/cell interfacial area, k_c, is defined as follows:

$$\begin{aligned} k_r C_m &= k_c a_m (C_{Li} - C_o) = k_m a_m (C_L - C_{Li}) \\ &= k_c a_m (C_{Li})_{cr} = k_m a_m [(C_L)_{cr} - (C_{Li})_{cr}]. \end{aligned} \tag{15}$$

Rearrangement of the above equation gives:

$$k_r C_m = \frac{(C_L)_{cr}}{\dfrac{1}{k_m a_m} + \dfrac{1}{k_c a_m}} \tag{16}$$

2.2 In Cultivation of Multicellular Organisms

In the Analysis of oxygen transfer in the cultivation of multicellular organisms, the oxygen transfer in the liquid films around air bubbles and mycelial pellets can be analyzed by the same method as above. The analysis of oxygen transfer in mycelial pellets is based on the following assumptions: 1. the mycelial pellets are spherical, 2. the oxygen transfer in a mycelial pellet is described by an equation of molecular diffusion and the diffusion coefficient, D_P, is the product of the molecular diffusion coefficient in culture medium and void fraction in mycelial pellet, ϵ, as follows:

$$D_P = D_L \cdot \epsilon \tag{17}$$

3. the specific oxygen-consumption rate of mycelial cells is given by a Michaelis-Menten type equation as follows:

$$k_r = \frac{(k_r)_{\max} C}{K_m + C}. \tag{18}$$

4. The mycelial density, ρ, in the pellet is uniform.

Taking the material balance for oxygen on a differential volume at a distance r from the center of the mycelial pellet in steady state on the above assumptions, the following equation can be derived:

$$\frac{d^2 C}{dr^2} + \frac{2}{r}\frac{dC}{dr} = \frac{\rho}{D_P}\left[\frac{(k_r)_{\max} C}{K_m + C}\right] \tag{19}$$

The oxygen-consumption rate per unit mycelial pellet, V_r', or per unit dry mycelial weight, k_r, is given by the following equation:

$$V_r' = 4\pi R^2 D_P \left(\frac{dC}{dr}\right)_{r=R} \quad \text{or} \quad k_r = \frac{3D_P}{R\rho}\left(\frac{dC}{dr}\right)_{r=R} \tag{20}$$

When the rate of oxygen penetration into the mycelial pellet is rapid and the oxygen concentration at the center of the mycelial pellet is equal to that at its surface, C_{Li}, the oxygen-consumption rate per unit mycelial pellet, V_r, is given by

$$V_r = \frac{4}{3}\pi R^3 \rho \frac{(k_r)_{\max} C_{Li}}{K_m + C_{Li}}. \tag{21}$$

The effectiveness factor, η, for oxygen-consumption rate per unit mycelial pellet is obtained by

$$\eta = \frac{V_r'}{V_r} = \frac{3D_P(k_m + C_{Li})}{R\rho(k_r)_{\max} C_{Li}} \left(\frac{dC}{dr}\right)_{r=R} \tag{22}$$

Equations (19) and (22) may be rearranged in dimensionless terms as follows:

$$\frac{d^2\gamma}{d\lambda^2} + \frac{2}{\lambda}\frac{d\gamma}{d\lambda} = \frac{\phi^2 \gamma}{1+\gamma} \tag{23}$$

$$\eta = \frac{3}{\phi^2}\left(\frac{1}{\gamma_s} + 1\right)\left(\frac{d\gamma}{d\lambda}\right)_{\lambda=1} \tag{24}$$

where

$$\begin{aligned} \phi &= R\sqrt{\rho(k_r)_{\max}/D_P K_m} \\ \gamma &= C/K_m, \qquad \gamma_s = C_{Li}/k_m \\ \lambda &= r/R. \end{aligned} \tag{25}$$

Boundary conditions are

$$\begin{aligned} \gamma &= \gamma_s \quad \text{at } \lambda = 1 \\ \frac{d\gamma}{d\lambda} &= 0 \quad \text{at } \lambda = 0. \end{aligned} \tag{26}$$

Equation (23) cannot be solved analytically, but it can be solved numerically by a computer and the effectiveness factor, η, may be obtained from Eq. (24), using the numerical solution of Eq. (23).

3. Oxygen Transfer in Cultivation of Single-Cell Organisms

It was reported by Aiba (1960), using the experimental results of Hixson and Gaden (1950) on the cultivation of baker's yeast, that the oxygen transfer rate per unit of liquid volume at the liquid film around cells was much higher than those at the liquid film around air bubbles and in the cells. It was found by Kobayashi *et al.* (1962), whilst discussing the experimental results of Terui *et al.* (1960a) on the cultivation of *Saccharo-*

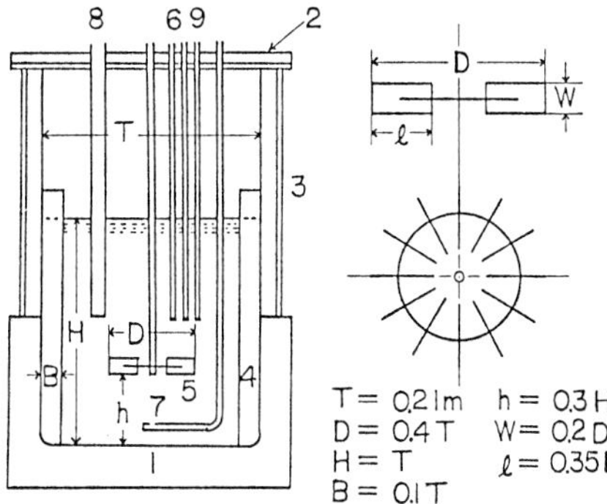

Fig. 1. Construction of jar fermenter and type of agitator used. 1) Jacket. 2) Cover-plate. 3) Tie rod. 4) Baffle. 5) Agitator. 6) Sampling tube. 7) Sparger. 8) Electrode for measurement of C_L. 9) Electrode for measurement of pH

myces cerevisiae, that the resistances to oxygen transfer in liquid films around cells and within the cells were much lower than those in liquid films around air bubbles.

Using a jar fermenter (see Fig. 1) and bakers' yeast, the oxygen transfer in the cultivation of bakers' yeast was investigated by the present author (Miura and Hirota, 1966) and the results are presented here. The composition of the defined medium used is shown in Table 1. Experiments were performed under the following conditions: agitation speed 220 to 400 R.P.M.; superficial gas velocity 6.1 to 27.5 m/hr with respect to the total cross-section of the tank; temperature 30° C; and pH 5.

Table 1. Synthetic medium composition for bakers' yeast

Glucose	10	g
Ammonium phosphate, monobasic	1	g
Potassium phosphate, monobasic	0.4	g
Magnesium sulfate	0.15	g
Sodium citrate	0.1	g
Biotin	0.02	mg
Calcium pantothenate	0.5	mg
Inosite	10	mg
Thiamine hydrochloride	4.4	mg
Pyridoxine hydrochloride	1.2	mg
Zinc sulfate	1.76	mg
Mohr's salt	1.05	mg
Cupric sulfate	0.096	mg
Distilled water	to 1 litre	
pH = 5: controlled by a concentrated aqueous solution of ammonia		

The volumetric oxygen-transfer coefficient for the liquid film around the air bubbles, k_La, was calculated by substituting observed values of k_r, C_m, C_L, and $dC_L/d\theta$ into Eq. (1). The oxygen-transfer coefficient for the liquid film around the air bubbles, k_L, was also assessed with reference to our previous work (Miura, 1961; Yoshida and Miura, 1963) on the gas/liquid interfacial area per unit of liquid volume, a, after correcting for the viscosity and surface tension of the culture medium; the correction allows for the fact that the gas/liquid interfacial area varies in inverse proportion to the 0.25 power of liquid viscosity and the 0.6 power of liquid surface tension, according to the experimental results of Vermeulen *et al.* (1955) and Calderbank (1958). The values of k_L are shown in Figs. 2, 3, and 4, in which the values of k_m and k_c estimated from Eqs. (12) and (16) are also shown. The values of k_m and k_L were of the same order of magnitude and these values were much greater than the values of k_c as shown in these figures. The actual values of k_m are considered to be greater than the values of k_L, since the values of k_m from Eq. (12) are expected to be smaller than the actual values.

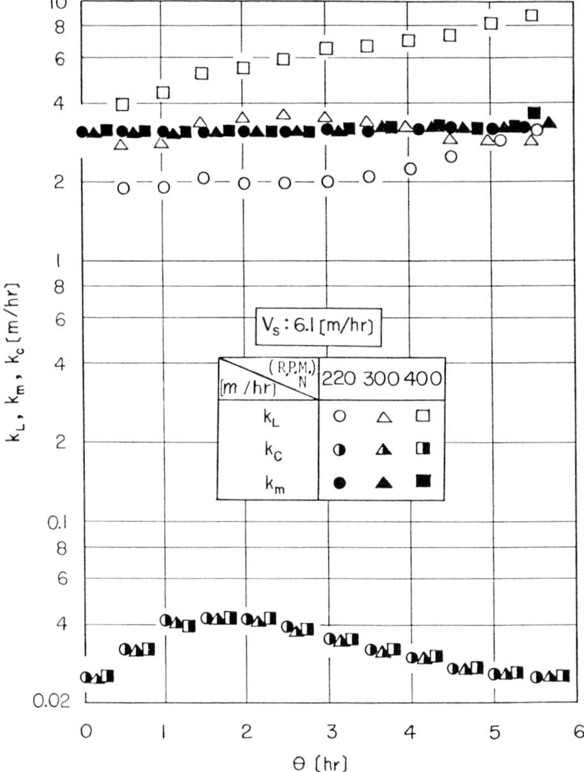

Fig. 2. k_L, k_m calculated by Equation (12) and k_c calculated by Equation (16) vs. fermentation time

The volumetric oxygen-transfer coefficients are shown in Figs. 5, 6, and 7. In most experiments, k_La values were less than k_ma_m and k_ca_m values as shown in these figures. The values of k_La and k_ca_m were nearly of the same order of magnitude, while k_ma_m values were much greater. These results suggest that the resistance per unit of liquid volume to oxygen transfer in the liquid films around air bubbles and the resistance to the oxygen uptake by the organism are much greater than the resistance in the liquid films around the cells. Since k_m values shown in Figs. 5, 6, and 7 were estimated by Eq. (12), these values were expected to be less than the actual values [cf. Eqs. (6) and (7)].

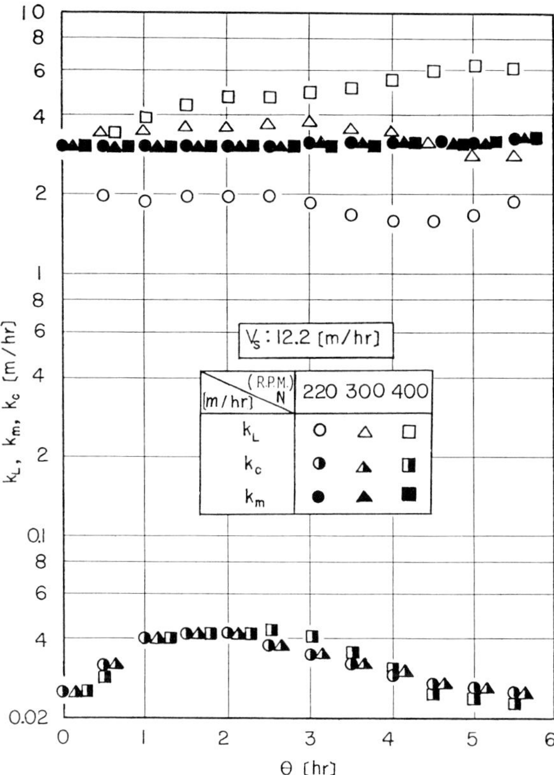

Fig. 3. k_L, k_m calculated by Equation (12) and k_c calculated by Equation (16) vs. fermentation time

If microbial cells from several microns to sub-microns in size are handled, the above argument will hold firm, suggesting that oxygen-transfer resistance in the liquid films around the cells can be neglected. Comparisons were made between $k_L a$ values calculated from the correlations for k_L and a presented in our previous work (Miura, 1961; Yoshida and Miura, 1963) and those in these experiments. The results showed reasonably good agreement as indicated in Fig. 8.

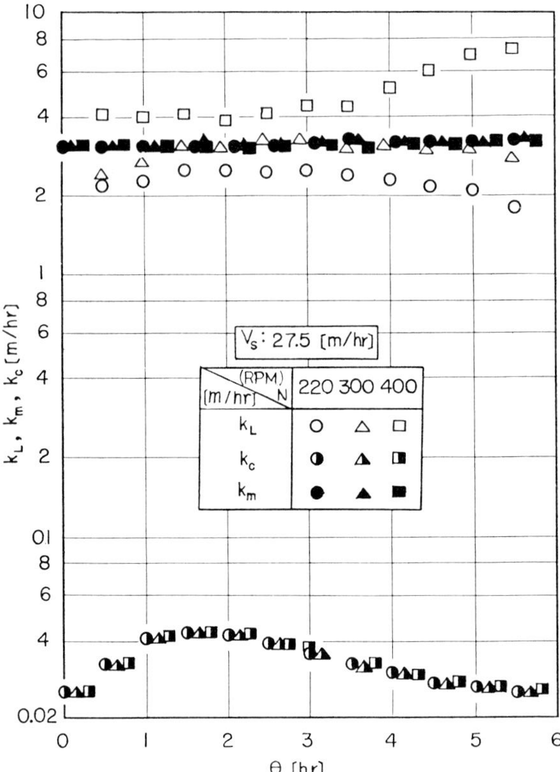

Fig. 4. k_L, k_m calculated by Equation (12) and k_c calculated by Equation (16) vs. fermentation time

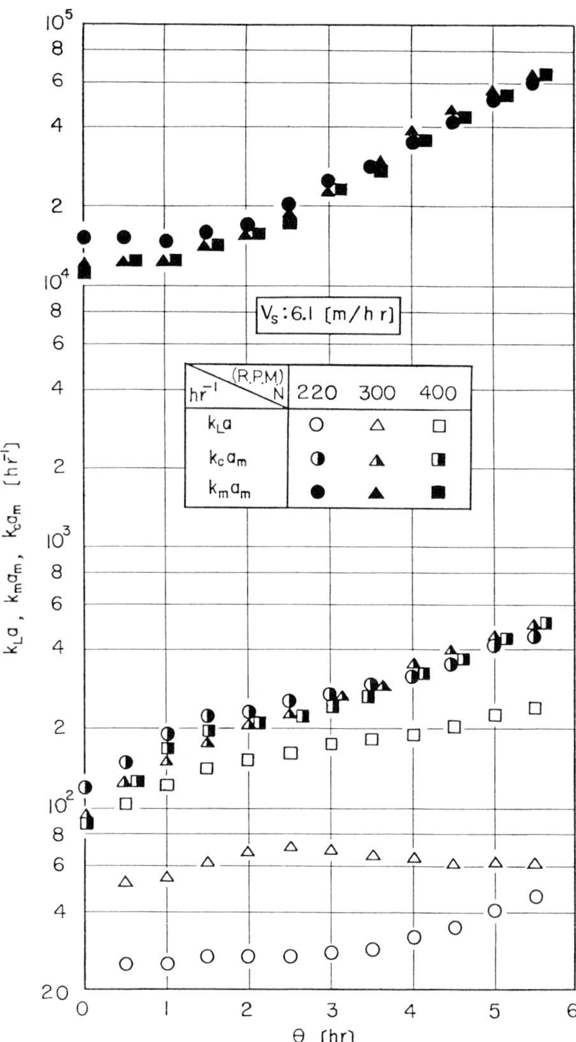

Fig. 5. k_La, k_ma_m and k_ca_m vs. fermentation time

4. Oxygen Transfer in Cultivation of Multicellular Organisms

4.1 Introduction

In the cultivation of multicellular or mycelial microbes, the formation of mycelial pellets and non-Newtonian character in the culture fluid are often observed. It is well known that the mycelial pellets may reduce the efficiency of fermentation, e.g., in penicillin production. It is, therefore, important, especially for the production of many antibiotics and other useful products, to discuss the mechanism of oxygen transfer in the cultivation of multicellular organisms.

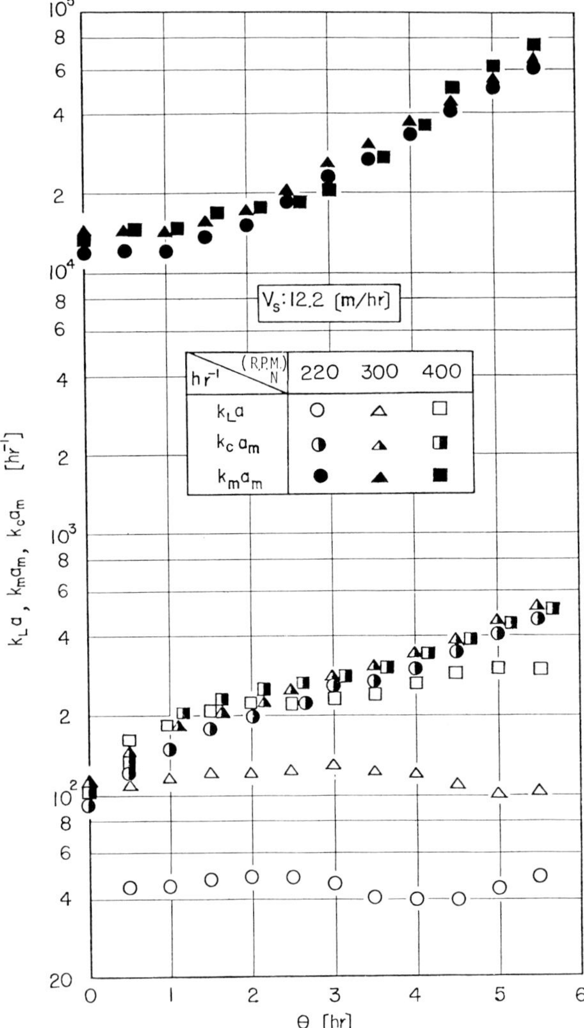

Fig. 6. k_La, k_ma_m and k_ca_m vs. fermentation time

Oxygen transfer within the mycelial pellet is considered to be the overall controlling step in oxygen-transfer processes in submerged fermentations with multicellular organisms. Bylinkina and Birukov (1972) showed that mass transfer from bulk liquid to pellets had to be accounted in the presence of large pellets. Yano *et al.* (1961), Yano and Yamada (1963), and Phillips (1966) investigated the diffusion of oxygen in a mycelial pellet and analyzed this problem on the assumption that the respiration reaction was of zero order with regard to the dissolved oxygen concentration. However, it was observed by Yoshida *et al.* (1967, 1968) and by the present author (Miura *et al.*, 1968) that the respiration reaction in mycelial pellets proceeded according to a Michaelis-Menten type equation and the respiratory activity varied, depending on the radial distance within the mycelial pellet; the respiration rate at the surface was higher than that at the center. The

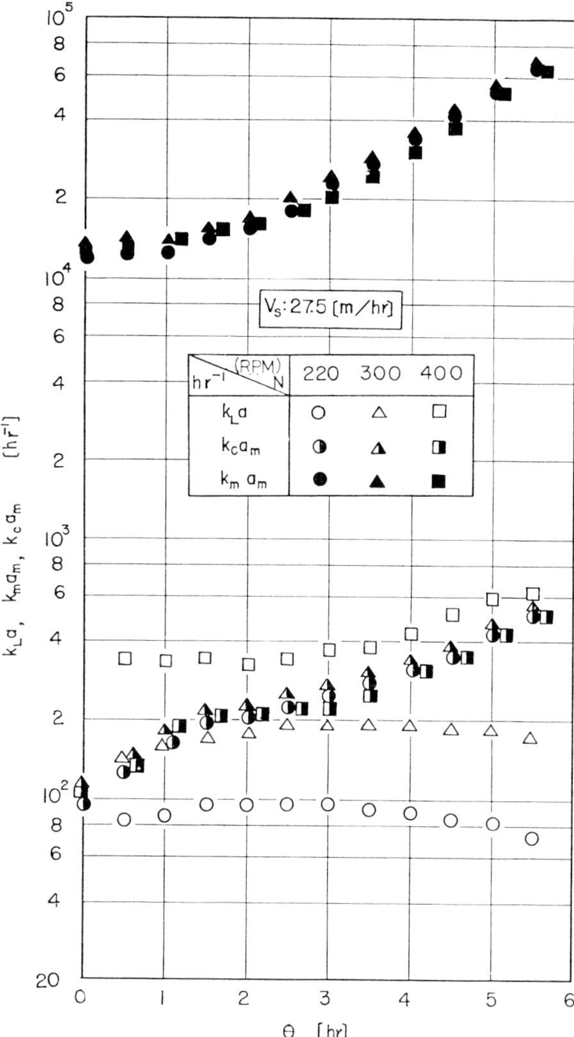

Fig. 7. k_La, k_ma_m and k_ca_m vs. fermentation time

results of investigations (Miura *et al.*, 1972, 1975) into oxygen transfer in the cultivation of multicellular organisms are presented below.

4.2 Formation of Mycelial Pellets

The organism used in our experiments was a strain of *Aspergillus niger* and the composition of the medium used is shown in Table 2. The organisms were precultured on an agar slant at 30° C for four to seven days. Spores formed on the slant were inoculated into a 500 ml Erlenmeyer flask containing 100 ml of the basal medium. Cultivation was carried out on a rotary shaker at a speed of 180 R.P.M. at 30° C for 46 to 52 hrs.

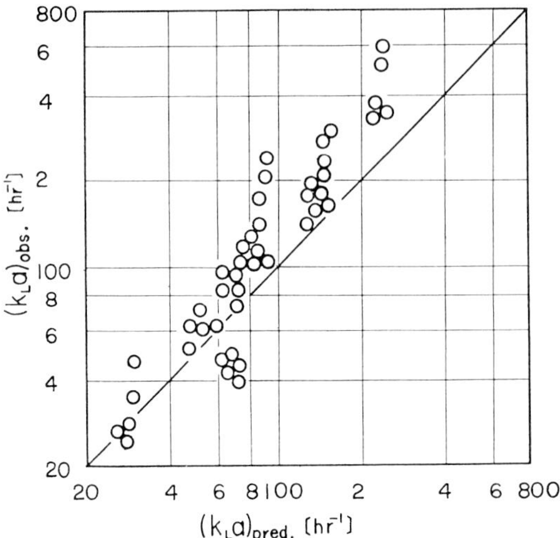

Fig. 8. Comparison of k_La between observation and calculation

Table 2. Composition of culture medium used for fungi

Glucose	50 g
Potassium phosphate, monobasic	1.0 g
Ammonium nitrate	0.8 g
Magnesium sulfate	0.5 g
Yeast extract	0.4 g
Distilled water	to 1 litre
pH = 6: controlled by an aqueous solution of sodium hydroxide.	

On investigation of the culture conditions necessary for a reproducible formation of the pellets, it was found that reproducibility was obtained under the above-mentioned conditions. Spherical mycelial pellets were formed by several other fungi under the same conditions and the mycelial densities of those pellets were not very different from *A. niger*, as shown in Table 3. The mycelial density of *A. niger* pellets was not changed much by the composition or viscosity of the culture medium. It is considered from the above results that the following results obtained from the pellets of *A. niger* are also applicable to some other fungi.

Table 3. Pellet formation and mycelial density of several fungi

Fungi	Formation of pellet	Density of pellet (g/cm^3)
Aspergillus niger	+	0.019
Aspergillus oryzae	+	0.023
Penicillium chrysogenum	+	0.051
Penicillium spiculisporam	+	0.057
Penicillium granulatum	+	0.040

4.3 Transfer Rate of Oxygen at Each Step of Its Transfer Processes

A glass jar, geometrically similar to the jar shown in Fig. 1, with a 1.4-litre capacity was used in this study. The values of $k_L a$ were assessed from Eq. (1). The oxygen-transfer coefficient for the liquid film around a mycelial pellet, k_m, was also estimated from Eq. (12), since the difference in the density between the mycelial pellet and the culture medium was not great and the mycelial pellets tend to follow the motion of the culture fluid. The terminal velocity of the mycelial pellet under gravity in the culture medium was observed to obtain the Peclet number. The results are shown in Fig. 9. It is clear

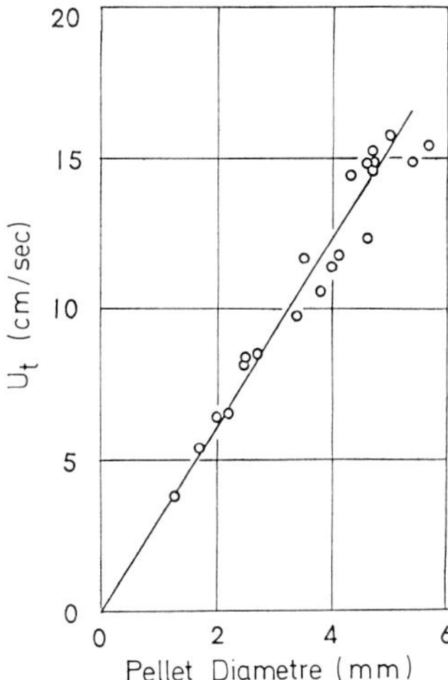

Fig. 9. Terminal velocity of mycelial pellet in free fall vs. diametre of mycelial pellet.

that the terminal velocity is proportional to the diameter of the mycelial pellet and the result is not in accordance with Stokes' law. The Peclet number was assessed using the values of U_t in Fig. 9. The oxygen-transfer coefficient for a mycelial pellet, k_p, is defined as follows:

$$\begin{aligned} k_r C_m &= k_p a_p (C_{Li} - C_o) = k_m a_p (C_L - C_{Li}) \\ &= k_p a_p (C_{Li})_{cr} = k_m a_p [(C_L)_{cr} - (C_{Li})_{cr}]. \end{aligned} \tag{27}$$

Accordingly, the k_p values were calculated from:

$$k_r C_m = \frac{(C_L)_{cr}}{\dfrac{1}{k_p a_p} + \dfrac{1}{k_m a_p}} \tag{28}$$

In the above equations, $(C_L)_{cr}$ is the critical oxygen concentration, implying that the mycelial respiration rate is independent of C_L, if C_L is greater than $(C_L)_{cr}$.
The relationship between the volumetric oxygen-transfer coefficient for liquid films around air bubbles, k_La, and the concentration of mycelium, C_m, is shown in Fig. 10, which indicates that k_La decreases more rapidly with increasing concentration of pulpy mycelium than with mycelial pellets.

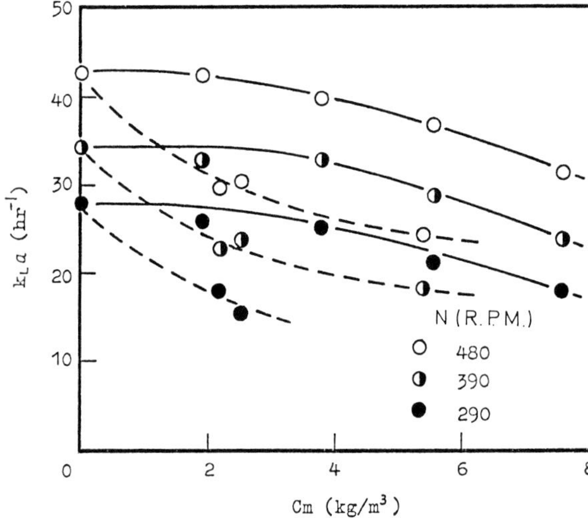

Fig. 10. Comparison of k_La in a suspension of pellets with that in a suspension of pulpy mycelia. —— mycelial pellets, – – – pulpy mycelia

The values of volumetric oxygen-transfer coefficients for liquid films around air bubbles and mycelial pellets, k_La, k_ma_p, and those for mycelial pellets, k_pa_p, are shown in Figs. 11 and 12. Incidentally, these values appear to be nearly of the same order of magnitude.

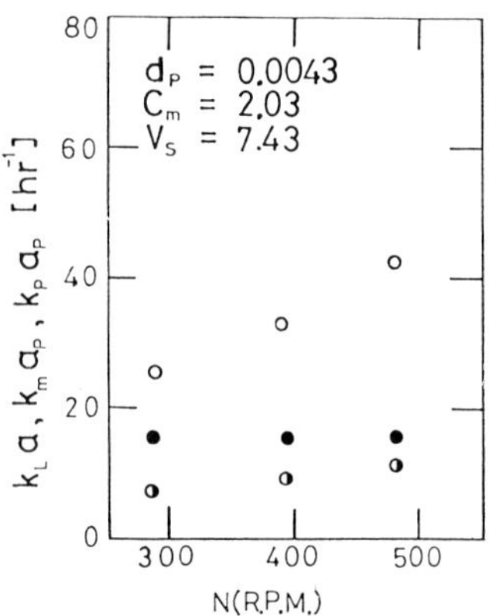

Fig. 11. k_La, k_ma_p and k_pa_p vs. agitation speed. ○: k_La, ●: k_ma_p, ◑: k_pa_p

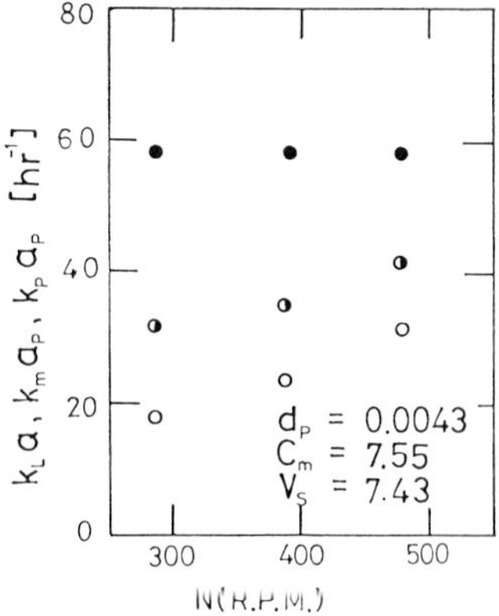

Fig. 12. k_La, k_ma_p and k_aa_p vs. agitation speed. ○: k_La, ●: k_ma_p, ◑: k_pa_p

The above treatment relies on the assumptions of stagnant liquid films around the air bubbles and mycelial pellets and of control of oxygen transfer by molecular diffusion. But further studies on the oxygen-transport phenomena with mycelial pellets suggest that oxygen transfer with the mycelial pellet is influenced by turbulence in the culture medium. Accordingly, the mechanism of oxygen transfer within mycelial pellets is discussed in the next section.

4.4 Oxygen Transfer within Mycelial Pellets

Previous discussions (Yano *et al.*, 1961; Yano and Yamada, 1963; Phillips, 1966; Yoshida *et al.*, 1967; Aiba and Kobayashi, 1971; Kobayashi *et al.*, 1973) about oxygen transfer within the mycelial pellet were based on the assumption that oxygen transfer was controlled by molecular diffusion. Miura *et al.* (1972, 1975) discussed the mechanism of oxygen transfer within the mycelial pellet, using mycelial pellets of *A. niger*, and it was found that the oxygen transfer within the mycelial pellet was influenced by agitation as shown below.

The mycelial pellets were separated from the culture broth and washed with sterilized water; the rate of oxygen uptake by the pellets was then assessed by measuring the decrease of the dissolved oxygen concentration in the nitrogen-free basal medium with an oxygen analyzer (Beckman Model 777) in an agitated closed vessel at 30° C. The oxygen-uptake rate per unit dry weight of mycelial pellet, k_r, was evaluated by measuring the dry weight of the pellets after measuring the oxygen-uptake rate. Experimental conditions are shown in Table 4.

Table 4. Experimental conditions

Variable	Range	
D	5, 6	(cm)
N	120 ~521	(R.P.M.)
d_p	0.13 ~ 0.52	(cm)
μ_L	0.48 ~ 6.1	(g/cm · min)
T	9	(cm)

Figure 13 shows the effects of agitation speed and dissolved oxygen concentration on the oxygen-uptake rate by the mycelial pellets. The chained line indicates the rate of oxygen uptake by the pulpy mycelia, obtained by carefully unravelling the mycelial pellet, and the broken line indicates the theoretical oxygen-uptake rate estimated by Eqs. (20), (21), and (24) on the assumption that the oxygen transfer in the pellet is controlled by molecular diffusion. The rate of oxygen uptake by the pellets was evidently increased by the agitation. The oxygen-uptake rates with agitation were higher than those with molecular diffusion. The resistance to oxygen diffusion in the oxygen uptake by pulpy mycelia is considered to be negligible compared with that by the pellets and the rate of oxygen uptake by pulpy mycelia was the highest.

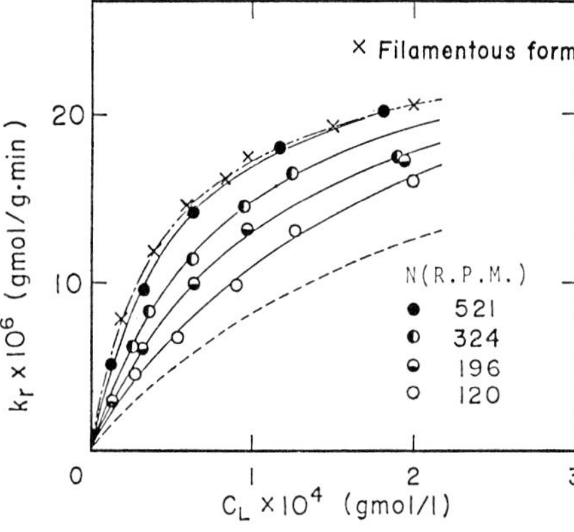

Fig. 13. Effects of agitation speed and dissolved oxygen concentration on oxygen-uptake rate of mycelial pellet. The chained line indicates oxygen-uptake rate of mycelia obtained by unravelling the pellets. The broken line shows the rate calculated on the assumption that oxygen is transferred by simple molecular diffusion

The experimental oxygen-uptake rates were evidently higher than the theoretical rates, as shown in Fig. 13. The difference between the experimental and theoretical values is considered to be caused by the following factors: (1) the oxygen transfer in the mycelial pellet is increased by agitation; (2) the mycelial density in the pellet is not uniform, for Miura *et al.* (1972) observed that the mycelial density at the surface was higher than that at the center of the comparatively large pellet of *A. niger*; (3) the respiratory activity is not uniform throughout the radial distance of the pellet, for Yoshida *et al.* (1968) and Miura *et al.* (1968) observed the non-uniformity of respiratory activity in large pellets of *Lentinus edodes* (24 mm in diameter) and *A. niger* (30 mm in diameter), whilst the theoretical oxygen-uptake rate in Fig. 13 was obtained on the assumption that $(k_r)_{max}$ and K_m were uniform and the same as those of pulpy mycelia. In industrial fermentations, small pellets below 1 mm in diameter are often formed. It is, therefore, considered that the non-uniformities of mycelial density and respiratory activity are low in those mycelial pellets often formed in industrial fermentations.

The increase of oxygen-uptake rates by agitation is considered to be caused by the following mechanisms: 1. the penetration of turbulence in the bulk medium into the mycelial pellet, 2. the deformation of the pellet by collision with the agitator or by the non-uniform rates of medium flow in the agitated fermenter. At present, it has not been verified which is the more important factor, 1. or 2., for the increase of oxygen-uptake rate by agitation. In this section, the mass transfer in the mycelial pellet was more quantitatively investigated on the assumption that the mass was transfered with an effective diffusivity, D_e, enhanced throughout the pellet by agitation.

The penetration rate of Blue Dextran into the mycelial pellet was measured in order to obtain the mass-transfer rate independent of physiological reactions, since the Blue Dextran cannot react with the mycelia and the adsorption of Blue Dextran on the mycelia is negligible.

Figure 14 shows the experimental results concerning the exudation of Blue Dextran from the pellets into the bulk solution on a reciprocal shaker. The results indicate only the effect of agitation speed on the rate of penetration of transfer material within mycelial pellets.

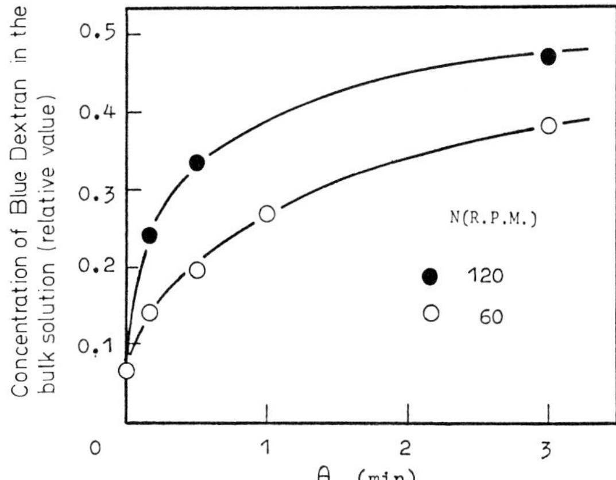

Fig. 14. Exudation of Blue Dextran into the bulk solution from the pellets at two agitation speeds in a 100 ml flask agitated by a reciprocal shaker

The rate of penetration of Blue Dextran into the pellets was measured by the following procedure: 1. a definite volume of the pellets was steeped in a Blue Dextran solution of high concentration in an agitated vessel with a 0.4 litre working volume, 2. the pellets were allowed to absorb the Blue Dextran during time θ, then screened off from the solution, 3. the pellets were mixed with a definite volume of distilled water on a reciprocal shaker at a speed of 120 R.P.M. for 3 hrs, 4. the quantity of Blue Dextran absorbed in the pellets was measured by analyzing the concentration of Blue Dextran in the aqueous solution suspending the pellets through a spectrophotometer at 630 mμ. According to Fick's law, the following equation describes the penetration phenomena with the diffusion coefficient D_e:

$$\frac{\partial C}{\partial \theta} = D_e \frac{\partial^2 C}{\partial x^2}. \tag{29}$$

Initial and boundary conditions are

$$\begin{array}{lll} \theta = 0, & x > 0, & C = \bar{C} \\ \theta > 0, & x = 0, & C = C_i \\ \theta > 0, & x = \infty & C = \bar{C}. \end{array} \tag{30}$$

The solution to Eq. (29) so as to satisfy boundary conditions of Eq. (30) yields

$$C = \bar{C} + \frac{C_i - \bar{C}}{\sqrt{\pi D\theta}} \int_x^{\infty} \exp.\left(-\frac{x^2}{4D\theta}\right) dx. \tag{31}$$

The mass flux rate at the interface, $x = 0$, can be derived from Eq. (31) according to Fick's law as follows:

$$f(\theta) = -D_e\left(\frac{\partial C}{\partial x}\right)_{x=0} = (C_i - \bar{C})\sqrt{\frac{D_e}{\pi\theta}}. \tag{32}$$

Therefore, the mass flux quantity, F, for time θ can be obtained by the following equation:

$$F = \int_0^{\theta} f(\theta) d\theta = 2(C_i - \bar{C}) \sqrt{\frac{D_e \theta}{\pi}} \tag{33}$$

The experimental results for Blue Dextran absorption into the mycelial pellet were compared with the above theoretical analysis and the effective diffusion coefficient of the transfer material within the mycelial pellet was estimated. The absorption of Blue Dextran into the pellets was carried out in an aqueous solution of Blue Dextran, the concentration of which was so high that the boundary condition of Eq. (30) was satisfied. These results, which are shown in Fig. 15, indicate that the flux quantities of Blue Dextran, F, are proportional to the square root of θ within this experimental period and the experimental flux quantities are much higher than the flux quantity by molecular diffusion, shown in the figure by a broken line. The flux quantity indicated by a broken line in Fig. 15 was obtained from Eq. (33) on the assumption that the diffusion coeffi-

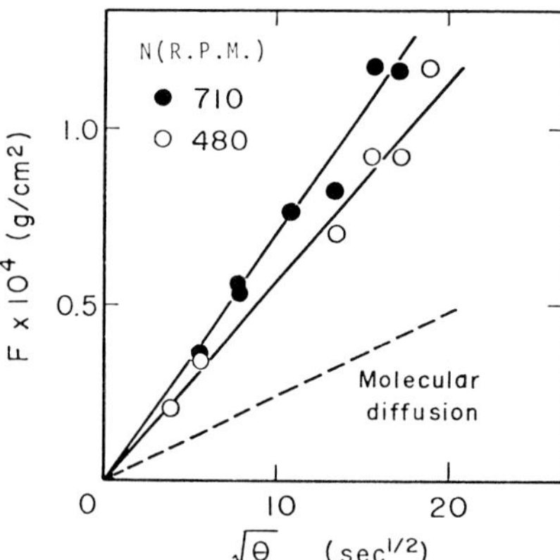

Fig. 15. Relation between F and $\sqrt{\theta}$ for penetration of Blue Dextran into pellets

cient of transfer material was equal to the molecular diffusion coefficient of oxygen in the mycelial pellet, D_p, estimated by Eq. (17). The molecular weight of Blue Dextran is much higher than that of oxygen and the diffusion coefficient of Blue Dextran will be much lower than that of oxygen. It is, therefore, considered that the quantity of Blue Dextran entering the mycelial pellet by molecular diffusion is lower than that indicated by the broken line in Fig. 15 and that the transfer of Blue Dextran in the mycelial pellet is highly influenced by agitation. The effective diffusion coefficient of Blue Dextran in the mycelial pellet was obtained by substituting the experimental results shown in Fig. 15 into Eq. (33) as follows: $D_e = 2.2 \times 10^{-4}$ cm^2/sec at an agitation speed of 710 R.P.M. and $D_e = 1.7 \times 10^{-4}$ cm^2/sec at an agitation speed of 480 R.P.M. The difference between the effective diffusion coefficients of oxygen and Blue Dextran is less in conditions influenced by turbulence than in conditions controlled by molecular diffusion. It is, therefore, considered that the effective diffusion coefficient of oxygen in the mycelial pellet is of the order of 10^{-4} cm²/sec or a little more, and it is greater by about one order of magnitude than its molecular diffusion coefficient under the conditions shown in Fig. 15.

Further, the oxygen uptake rates by the mycelial pellets were theoretically predicted and the results were compared with the experimental findings. The analysis of oxygen transfer in the mycelial pellets was made on the following assumptions: 1. the pellets were spherical; 2. the oxygen transfer in the pellet was accomplished with an effective diffusion coefficient, D_e, enhanced by agitation; 3. the oxygen-consumption rate of mycelial cells was given by a Michaelis-Menten type equation, according to the experimental results shown in Fig. 17; and 4. the physiological activity and mycelial density, ρ, were uniform in the pellet. Equations (18) to (26) apply on the above assumptions, using an effective diffusion coefficient of dissolved oxygen, D_e, instead of a molecular diffusion coefficient in the pellet, D_p.

The oxygen-uptake rate per unit mycelial pellet, V_r', is calculated by Eqs. (21) and (24), using D_e instead of D_p. The values of k_r were obtained by Eq. (20) from values of V_r' and the dry weight of unit mycelial pellet. The results are shown in Fig. 16, in which the values of D_e varied from the molecular diffusion coefficient of oxygen to infinity. The comparison between calculated and observed values in Fig. 16 shows that the above-mentioned model can predict the rate of oxygen transfer in the mycelial pellet.

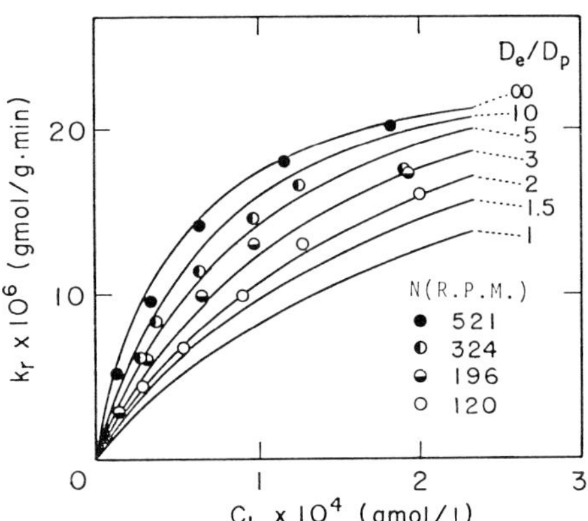

Fig. 16. Comparison between calculated and observed relations between k_r and C_L

Oxygen transfer within pellets has been studied on the assumption that the pellets behaved in much the same manner as a porous rigid sphere and that oxygen was transfered by simple molecular diffusion. Recently, Huang and Bungay (1973) measured oxygen concentrations in and near mycelial pellets and estimated oxygen diffusivity within pellets. The oxygen diffusivity obtained was 2.9×10^{-6} cm²/sec and it was much smaller than the molecular diffusivity of oxygen in water (1.8×10^{-5} cm²/sec at 20° C). Their measurements were, however, carried out for a fixed pellet in a slow-flowing medium (28.3 cm/min). The situation may be different from that in fermentations with vigorous agitation. Actually, the effective diffusivities within the pellets evaluated in our present study were more than twice the molecular diffusivity of oxygen. Accordingly, the oxygen-transfer rate in the liquid phase around the mycelial pellet is considered to be much higher than its molecular diffusion rate; k_m is much higher than k_L and k_P.

4.5 Factors Affecting the Rate of Oxygen Uptake by Mycelial Pellets

It is inferred from the results shown in Fig. 13 that the rate of oxygen uptake by the mycelial pellet is expressed by a Michaelis-Menten type equation:

$$k_r = \frac{(k_r)_{\max} C_L}{K'_m + C_L}. \tag{34}$$

The maximum oxygen-uptake rate, $(k_r)_{\max}$, and the apparent Michaelis-Menten constant, K'_m, can be obtained on the basis of the method of Terui and Konno (1960b). The relation between the initial oxygen concentration, $(C_L)_i$, and the half-life period, $t_{1/2}$, can be expressed by the following equation through integrating Eq. (34),

$$(C_L)_i = 2(k_r)_{\max} \cdot t_{1/2} - 1.38\, K'_m. \tag{35}$$

The relation between $(C_L)_i$ and $t_{1/2}$ was experimentally investigated and the results, which are shown in Fig. 17, indicate that $t_{1/2}$ is linearly related with $(C_L)_i$ as expressed by Eq. (35). The parameters, $(k_r)_{\max}$ and K'_m, can be obtained from the slope of the

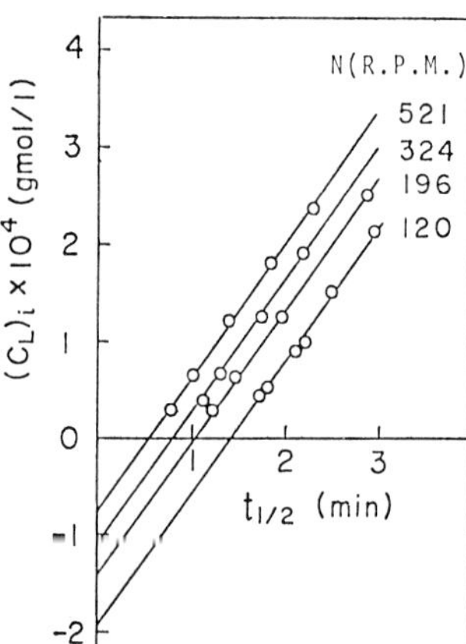

Fig. 17. Relation between $(C_L)_i$ and $t_{1/2}$ for oxygen uptake by mycelial pellets

straight line and the intersection of the straight line and ordinate in Fig. 17. It was found that the values of K_m' changed with the agitation speed, while the values of $(k_r)_{max}$ were approximately constant.

The oxygen uptake by the mycelial pellet is expressed by the following equation when the oxygen concentration is much lower than the value of K_m',

$$k_r = \frac{(k_r)_{max}}{K_m'} \cdot C_L \tag{36}$$

where $(k_r)_{max}/K_m'$ is the apparent first order reaction rate coefficient. The effects of various operation conditions upon $(k_r)_{max}/K_m'$ were investigated in relation to the rate of oxygen uptake by the pellet. The oxygen-uptake rates were measured for agitator diameter, D, agitation speed, N, pellet diameter, d_p, and liquid viscosity, μ_L, of ranges shown in Table 4. The results are shown in Fig. 18, which indicates that the oxygen-

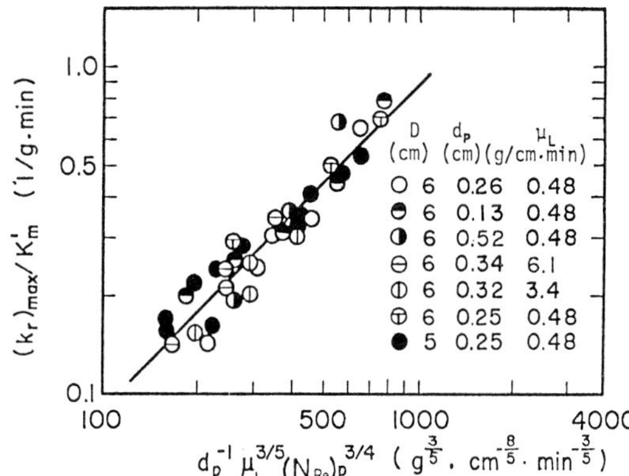

Fig. 18. Correlation for oxygen-uptake rates of mycelial pellets in unbaffled vessel

uptake rate is correlated with the operation parameters by the following equation within the operation conditions shown in Table 4,

$$(k_r)_{max}/K_m' \propto d_p^{-1}\mu_L^{3/5}(N_{Re})_p{}^{3/4}. \tag{37}$$

Figure 19 shows the experimental results of the oxygen-uptake rate in the same agitated tank equipped with four full baffles. The broken line in Fig. 19 is identical to the solid line in Fig. 18. The rate of oxygen uptake by the mycelial pellet was increased by the baffles.

A correlation could be developed for the rate of oxygen uptake by the mycelial pellets in this work. A more general correlation, applicable to morphologically similar pellets of other fungi, may be obtained by investigating the effects of operation conditions on the effective oxygen diffusivity, D_e, instead of the oxygen-uptake rate, since the value of D_e is independent of respiratory activity of the mycelial pellet.

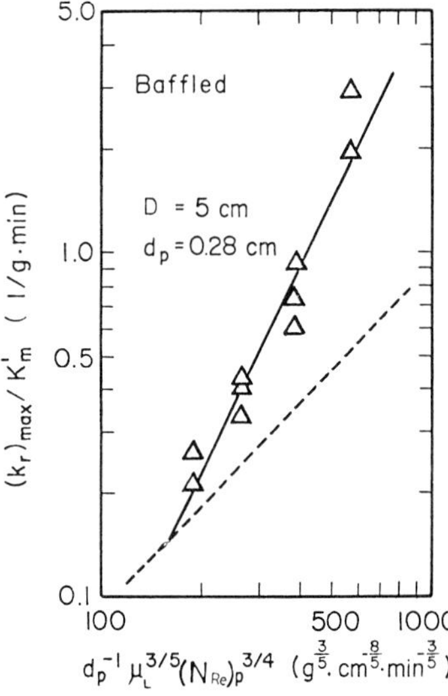

Fig. 19. Correlation for oxygen-uptake rates of mycelial pellets in baffled vessel

5. Scale-Up of Submerged Aerobic Fermenters

5.1 Criteria of Scale-Up

Submerged aerobic fermentation involves many factors which must be taken into account in scale-up; mass and heat transfer between gas, liquid, and cell; mixing of gas, liquid, and cell; dispersions of bubbles, cells, and liquid drops in heterogeneous fermentation; effect of shear by agitation on microorganisms, etc. Therefore, a general method for scale-up has not been found. At present, scale-up is carried out practically on the criterion of agitation power consumed per unit medium volume or volumetric oxygen rate. Scale-up on the criterion of oxygen-transfer rate is considered to be more reasonable than that on the criterion of agitation power consumed, although the latter is more practical than the former. The necessary agitation power obtained by the latter method is generally higher than that obtained by the former method. The mechanism of the oxygen transfer discussed above is the basis of the scale-up by the former method.

5.2 Scale-Up on the Criterion of Agitation Power Consumed

The correlation between the concentration of penicillin produced and agitation power consumed was investigated in fermenters of different volumes by Bartholomew *et al.* (1950), Gaden (1961), and Humphrey (1964a,b) and they reported that the scale-up of penicillin production could be successfully performed at an agitation power of 1.5 to 3.0 HP/m^3. It was reported by Wegrich *et al.* (1953) that the titre of penicillin produced in a 2000-gallon fermenter was reproduced in a 24000-gallon fermenter with the same values of superficial gas velocity and agitation power consumed per unit liquid volume. According to Bartholomew *et al.* (1950) and Humphrey (1964a, b), the scale-up of streptomycin production could be successfully performed with an agitation power of about 2 HP/m^3.

Agitation power in a stirred tank has been studied by many investigators. Rushton *et al.* (1950) obtained the following correlation for agitation power without aeration:

$$\frac{P_0 g_c}{\rho_L N^3 D^5} = K_1 \left(\frac{D^2 N \rho_L}{\mu_L}\right)^m \left(\frac{DN^2}{g}\right)^n \tag{38}$$

$$N_p = K_1 (N_{Re})^m (N_{Fr})^n .$$

In agitated tanks with full baffles, the vortex flow of liquid is not developed, the Froud number, N_{Fr}, is negligible and the above relation becomes:

$$N_p = K_2 (N_{Re})^m \tag{39}$$

K_1 and K_2 = coefficients

where the exponent, m, of Reynolds number was equal to -1 in laminar flow, from 0 to -1 in the transient state, and 0 in turbulent flow. Oyama *et al.* (1955) found experimentally that the correlation between the agitation power with aeration and that without aeration was expressed by the following equation;

$$P_g/P_0 = f(q/ND^2W) \tag{40}$$

In geometrically similar tanks

$$P_g/P_0 = f(q/ND^3) \tag{41}$$

The above correlation was changed by the type and number of agitators. Calderbank (1958) and Kalinske (1955) obtained a similar correlation to that of Oyama *et al.* (1955) shown above. Michel and Miller (1962) proposed the following correlation by which their own and others' experimental data of agitation power with aeration were correlated in wide ranges of liquid viscosity:

$$P_g = c(P_0^2 ND^3/q^{0.56})^{0.45} \tag{42}$$

where c = coefficient

Nagata *et al.* (1956) obtained empirical equations for the power consumption of paddle agitators having arbitrary dimensions in a vessel containing liquids of various viscosities. In the case of a concentric agitator without baffles,

$$\frac{P_0 g_c}{\mu_L N^2 D^3} = 310 \left(\frac{W}{T}\right)\left(\frac{D}{T}\right)^{1.15}, \text{ at } N_{Re} < 10 \text{ to } 10^2 \tag{43}$$

$$\frac{P_0 g_c}{\rho_L N^3 D^5} = A \cdot B \left(\frac{D^2 N \rho_L}{\mu_L}\right)^p \quad \text{at } 10 \text{ to } 10^2 < N_{Re} < 10^4 \text{ to } 10^5 \tag{44}$$

where
$$A = 10^{\left[\frac{W/T}{0.03 + 0.42(W/T)}\right]}$$

$$B = 10^{-\left[\frac{(W/T)(D/T)}{0.01 + 0.73(W/T)}\right]} \tag{45}$$

$$P = -\left[\frac{(W/T)}{0.17 + \{1.8 + 1.37(D/T)\}(W/T)}\right]$$

$$\frac{P_0 g_c}{\rho_L N^3 D^5} = \frac{(W/T)}{0.05 + 1.05(W/T)} \left(\frac{D}{T}\right)^{-1.23} \left(\frac{D^2 N \rho_L}{\mu_L}\right)^{-0.113} \tag{46}$$

at $N_{Re} = 10^4$ to 10^5.

In the fully baffled condition, power consumption was independent of liquid viscosity and liquid depth, and could be obtained by the following equations:

$$\frac{P_{max} g_c}{\rho_L N^3 D^5} = 23 \left(\frac{W}{T}\right)^{1.27} \left(\frac{D}{T}\right)^{-1} \tag{47}$$

at $W = 0.05T$ to $0.2T$, $D = 0.3T$ to $0.8T$

$$\frac{P_{max} g_c}{\rho_L N^3 D^5} = 12 \left(\frac{W}{T}\right)^{0.85} \left(\frac{D}{T}\right)^{-1} \tag{48}$$

at $W = 0.2T$ to $0.4T$, $D = 0.3T$ to $0.7T$.

Nagata *et al.* (1957) obtained the following experimental equation which was applicable for the power consumption of paddle-agitators in all the above-mentioned ranges with slight errors:

$$\frac{P_0 g_c}{\rho_L N^3 D^5} = \frac{A}{N_{Re}} + B \left(\frac{10^3 + 1.2 N_{Re}^{0.66}}{10^3 + 3.2 N_{Re}^{0.66}}\right)^p \left(\frac{H}{T}\right)^{(0.35 + W/T)} (\sin\theta)^{1.2} \tag{49}$$

where $A = 14 + (W/T)\{670(D/T - 0.6)^2 + 185\}$

$$B = 10^{\{1.3 - 4(W/T - 0.5)^2 - 1.14(D/T)\}} \tag{50}$$

$$P = 1.1 + 4(W/T) - 2.5(D/T - 0.5)^2 - 7(W/T)^4$$

θ = angle of blades to the horizontal plane

Nagata *et al.* (1967) presented the following empirical equation for power consumption by agitation accompanied by aeration over wide ranges of aeration rate, liquid viscosity, and ratio of impeller diameter to tank diameter:

$$\log \frac{P_g}{P_0} = -192 \left(\frac{D}{T}\right)^{4.38} \left(\frac{D^2 N \rho_L}{\mu_L}\right)^{0.115} \left(\frac{DN^2}{g}\right)^{1.96(D/T)} \left(\frac{q}{ND^3}\right) \tag{51}$$

Calderbank *et al.* (1959a) investigated the agitation power in non-Newtonian fluids and obtained the correlation between power number, N_p, and modified Reynolds number expressed by the following equation:

$$\frac{D^2 N \rho_L}{\mu_L} = \left(\frac{D^2 N^{2-n} \rho_L}{0.1K}\right)\left(\frac{n}{6n+2}\right) \tag{52}$$

where K and n were a coefficient and a power in the equation:

$$\tau = K\left(\frac{du}{dr}\right)^n \tag{53}$$

where u = fluid velocity, cm/sec,
r = distance perpendicular to fluid flow direction, cm
τ = shear stress, dynes/cm^2.

Taguchi and Miyamoto (1966) observed that P_g/P_0 depended not only upon the Reynolds number but also upon the rheological properties of the fluid in the agitation of non-Newtonian fermentation broths with and without aeration. In addition, they found that the correlation developed by Michel and Miller (1962) was applicable to non-Newtonian fluids in the turbulent region but it did not apply to non-Newtonian fluids in laminar and transient regions, and the impeller diameter and impeller blade width had a considerable effect on power consumption in nongassed non-Newtonian fluids.

5.3 Scale-Up on the Criterion of Oxygen-Transfer Rate

The scale-up of aerobic fermenters has often been successfully carried out on the criterion of the oxygen-transfer rate. The limiting steps for oxygen transfer are considered to be located in the liquid films around air bubbles and in cells in the cultivation of yeasts or bacteria in aerated agitated tanks as discussed above, while the resistance to intracellular oxygen transfer is negligible compared with that to oxygen transfer in the liquid films around air bubbles for microorganisms of high respiration rate as shown by Kobayashi (1962). It is, therefore, reasonable that scale-up is based on the oxygen-transfer rate in the liquid films around air bubbles for the cultivation of yeasts or bacteria. In the cultivation of multicellular organisms, the limiting steps for the oxygen-transfer rate are considered to be located in the liquid films around air bubbles and in mycelial pellets as discussed above and, therefore, the scale-up should be performed on the basis of these criteria.

Many authors, e.g. Strohm *et al.* (1959), Roxburgh *et al.* (1954), Bartholomew *et al.* (1960), and Richards (1961), have found that $k_L a$ determined by the sulphite method could be applied to the scaling-up of various fermentations. Bartholomew *et al.* (1960) related $k_L a$ to operating variables by the equation:

$$k_L a = k(P_g/V)^{0.4}(V_s)^{0.5}(N)^{0.5} \quad (54)$$

where k = coefficient,

whilst Fukuda *et al.* (1968) found that the performance of multiple impellers could be correlated by the equation:

$$k_d = (2.0 + 2.8N_i)(P_g/V)^{0.56} V_s^{0.7} N^{0.7} \times 10^{-3} \quad (55)$$

where k_d = volumetric oxygen-transfer coefficient (10^{-6} g. mole O_2/ml-min-atm.)
N_i = number of impellers,
P_g = power consumption accompanied by aeration, HP
V_s = superficial gas velocity based on cross section of tank, cm/min
N = agitation speed, R.P.M.

Blakebrough and Sambamurthy (1966) used the sulphite oxidation method to assess the mass-transfer performance of a wide range of turbine impellers in two-phase and three-phase systems, expressing their results by Eqs. (56) and (57):

$$k_L a = C_1\{t^{0.203}(P_g/V)^{1.79} M_f^{-1.05} N^{-0.0459}\} \quad (56)$$

$$\frac{k_L a}{N} = C_2\left[\frac{T^2H}{WL(D-W)}\right]^{1.437}\left[\frac{1}{N_t}\right]^{1.087}\left[\frac{D}{T}\right]^{1.02} \quad (57)$$

where M_f = momentum factor
= $ND \cdot NWL(D-W)$
D = impeller diameter, m
W = impeller blade width, m
L = impeller blade height, m
T = vessel diameter, m
H = height of suspension, m
C_1, C_2 = coefficients
t = mixing time, hr.

Equation (57) is dimensionless and may be used for scaling-up purposes.

But the scale-up of aerobic fermentations cannot always be carried out on the basis of sulfite oxidation rate. For example, Maxon and Steel (1959, 1962, 1966a) reported that the scale-up of the novobiocin fermentation could not be performed on the basis of sulfite oxidation rate, but that the oxidation availability rate for *Streptomyces niveus* could be the criterion for the scale-up of novobiocin production, the oxygen availability rate being correlated with the impeller tip velocity, independent of the diameter for the five turbine sizes examined. Steel and Maxon (1966b) compared the performance of a multiple-rod mixing impeller with that of conventional turbine impellers in viscous novobiocin beers. They found that the power requirement was independent of changes in apparent viscosity of the fermentation beer and the multiple-rod impeller gave the same novobiocin yield and oxygen availability rate at about one-half of the power required by turbines. Robinson and Wilke (1972) also observed that oxygen-transfer rates in electrolyte solutions such as sodium sulfate and sulfite solutions were higher than that in water and they presented a generalized correlation for oxygen transfer, taking into account the ionic strength in the solution.

The mechanism of the sulfite oxidation reaction was discussed by Yoshida *et al.* (1960) and Miura (1961) and according to them, the oxygen absorption rate for aqueous sodium sulfite solutions catalyzed by cupric ions was controlled by mass transfer in the liquid phase because, when gas/liquid interfacial areas were equal, absorption rates in pure water and both sodium sulfate and sulfite solutions were identical. They showed further that, in bubbling gas/liquid contactors with mechanical agitation operating under the same conditions, the oxygen absorption rate in sodium sulfate solution per unit volume of liquid was faster than in pure water, especially under high agitation and that this difference resulted from smaller bubbles and greater gas/liquid interfacial area in the sodium sulfite solution rather than from chemical reaction. It is, therefore, considered that the sulfite oxidation method is not suitable for the scale-up of aerobic fermentation, especially in highly viscous or non-Newtonian culture fluids or under high agitation.

In cultivations of mycelial-form organisms, the culture media are generally highly viscous or non-Newtonian and then the oxygen transfer is different from that in cultures of bacteria or yeasts as shown in the above section. Not only the oxygen transfer in the liquid phase around air bubbles but also that in the mycelial pellet must be taken into account in the cultivation of mycelial-form organisms as discussed above. Therefore, the scale-up of these fermentations should be carried out on the criteria of oxygen transfer in the liquid film around the gas-bubbles and in the mycelial pellets, using the results shown in the above section.

Mass transfer between gas and liquid in aerated, agitated gas/liquid contactors has also been studied by other investigators. Cooper *et al.* (1944) proposed the following correlation for oxygen absorption by sodium sulfite solutions in stirred gas/liquid contactors:

$$K_v = k(P_g/V)^{0.95} V_s^{0.67} \tag{58}$$

where K_v = overall volumetric oxygen transfer coefficient
k = coefficient.

Friedman and Lightfoot (1957) obtained experimentally the following correlation by the sulfite oxidation method:

$$K_v = CN^{3.08}D^{4.94} \tag{59}$$

where C = coefficient.
Yoshida *et al.* (1960) and Miura (1961) obtained the following correlations for oxygen absorption by water in three agitated tanks of different sizes:

$$k_L a = cV_s^m(N^3T^2)^n \tag{60}$$

For a vaned-disk agitator,

V_s, ft/hr	cV_s^m	n
5.4	2.1×10^{-4}	0.68
10	7.0×10^{-4}	0.63
20	2.6×10^{-3}	0.58
40	1.3×10^{-2}	0.50
90	6.2×10^{-2}	0.43

and the values of m were related to N^3T^2 by

N^3T^2	10^6	10^7	10^8
m	0.84	0.60	0.40

where N = agitation speed, R.P.M.
T = diameter of tank, ft.

For a turbine agitator, c was equal to 1.10 and m and n were each equal to 2/3. N^3T^2 is directly proportional to the power expended per unit volume. Values obtained for n are smaller than the value of 0.95 reported (Cooper *et al.*, 1944) for sulfite oxidation with a vaned-disk agitator. It is also less than the value of unity obtained for sulfite oxidation with flat paddle agitators (Friedman and Lightfoot, 1957).
In order to obtain more general correlations between mass transfer and the design and operation variables, k_L should be separated from the interfacial area per unit volume, a, because k_L and a are each influenced by different factors. Calderbank *et al.* (1958, 1959b, 1961) measured k_La by absorption of O_2, CO_2, SO_2, and CS_2 gases into water and aqueous solutions of glycol, alcohol, and glycerol in stirred tanks and separated k_L from a measured by the light-reflection or light-transmission techniques and obtained the following correlations for a and k_L:

$$\text{at } \left(\frac{D^2N\rho_L}{\mu_L}\right)^{0.7}\left(\frac{ND}{V_s}\right)^{0.3} < 20\,000$$

$$a = 1.44 \left[\frac{(P_g/V)^{0.4} \rho_L^{0.2}}{\sigma^{0.6}}\right] \left(\frac{V_s}{V_t}\right)^{1/2} \quad (61)$$

where a = gas/liquid interfacial area per unit liquid volume, cm^2/cm^3
P_g = power consumption accompanied by aeration, HP
V = liquid volume in tank, ft'^3
ρ_L = liquid density, gr/cm^3
σ = surface tension, dynes/cm

at $\left(\frac{D^2 N \rho_L}{\mu_L}\right)^{0.7} \left(\frac{ND}{V_s}\right)^{0.3} > 20\,000$

$$\log_{10} \frac{2.3a}{a_0} = 1.95 \times 10^{-5} \left(\frac{D^2 N \rho_L}{\mu_L}\right)^{0.7} \left(\frac{ND}{V_s}\right)^{0.3} \quad (62)$$

where a_0 = gas/liquid interfacial area per unit liquid volume for low Reynolds numbers, cm^2/cm^3

for $d_{BM} < 0.8$ mm

$$k_L \left(\frac{\mu_L}{\rho_L D_L}\right)^{2/3} = 0.31 \left(\frac{\Delta\rho\mu_L g}{\rho_L^2}\right)^{1/3} \quad (63)$$

for $d_{BM} > 2.5$ mm

$$k_L \left(\frac{\mu_L}{\rho_L D_L}\right)^{1/2} = 0.42 \left(\frac{\Delta\rho\mu_L g}{\rho_L^2}\right)^{1/3} \quad (64)$$

When the particles of the dispersed phase are not free to move under gravity and transfer is due to turbulence in the surrounding liquid,

$$k_L \left(\frac{\mu_L}{\rho_L D_L}\right)^{2/3} = 0.13 \left[\frac{(P_g/V)\mu_L}{\rho_L^2}\right]^{1/4} \quad (65)$$

where P_g = power consumption accompanied by aeration, kg m^2/hr^3.
Miura (1961) and Yoshida and Miura (1963) measured the gas/liquid interfacial area by a method utilizing the rate of gas absorption with a moderately fast first order reaction. The average bubble diameter was calculated from the interfacial area and the observed values of the gas hold-up determinde by measuring the bulk density of the aerated liquid, resulting in the following correlations:

with a turbine agitator

$$
\begin{aligned}
a &\propto (ND)^{1.1} V_s^{0.75} \\
H_G &\propto N^{0.8} D^{1.2} V_s^{0.75} \\
d_{BM} &\propto N^{-3} D^{0.1}
\end{aligned}
\tag{66}
$$

with a vaned-disk agitator

	V_s, ft/hr		
	20	40	90
a	$\propto (ND)^{0.9}$	$(ND)^{0.8}$	$(ND)^{0.7}$
H_G	$\propto N^{0.8}D$	$N^{0.7}D$	$N^{0.6}D$
d_{BM}	$\propto N^{-0.1}D^{0.1}$	$N^{-0.1}D^{0.2}$	$N^{-0.1}D^{0.3}$

Then, values of k_L were obtained from the data on k_La mentioned above (Yoshida *et al.*, 1960; Miura 1961) and the present correlations for a. These data on k_La included data for oxygen absorption by water at 7°, 20°, and 30° C and by aqueous solution of glycerol at 15° C with a range of viscosity from 1.9 to 3.6 cp and 20-fold variation in the Schmidt number. Fig. 20 is a log-log plot for all the runs with three

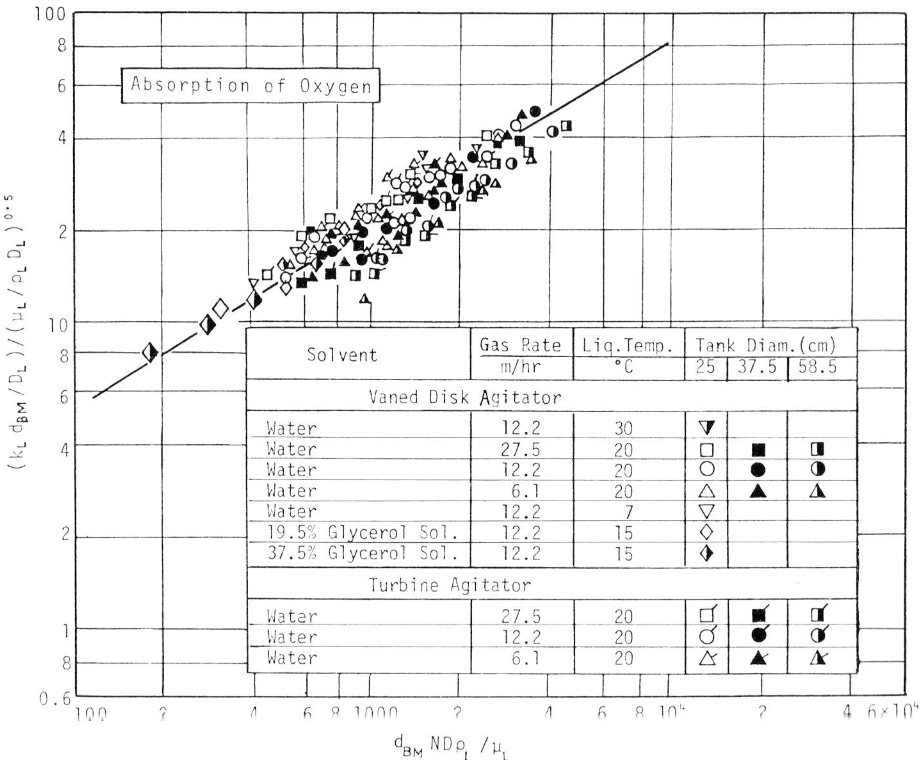

Fig. 20. General correlation for k_L in agitated gas liquid contactors

sizes of tanks and two types of agitators. The straight line through the data points is expressed by the equation:

$$\frac{k_L d_{BM}}{D_L} = 0.33 \left(\frac{d_{BM} N D \rho_L}{\mu_L}\right)^{0.6} \left(\frac{\mu_L}{\rho_L D_L}\right)^{0.5} \quad (67)$$

The agreement of the data for the two types of agitators might be a coincidence, because the value of the constant in the above equation should vary with the geometry of equipment.

According to Calderbank *et al.* (1961), k_L was proportional to the 1/4 power of the agitator power consumed per unit liquid volume. Since the power required for agitation is approximately proportional to the cube of agitation speed, this result implies that k_L varies with the 3/4 power of the agitation speed, which is in slight disagreement with our correlation. They also stated that the type of agitator used may have a minor influence on k_L for a given value of the agitation power dissipation per unit volume. Values of k_L obtained in our work range from 0.6 to 3.4 m/hr are somewhat higher than values calculated by the Calderbank correlation. The values of $k_L a$ predicted from the correlations for k_L and a presented by us agreed approximately with those measured in the cultivation of bakers' yeast as shown in Section 3.

Westerterp *et al.* (1963) also measured the gas/liquid interfacial area by chemical absorption and obtained the following correlations:

$$aH/H_L = C(N - N_0) D \left(\frac{\rho_L T}{\sigma}\right)^{1/2} \quad (68)$$

where $C = (0.79 \pm 0.16)\mu_L$
$\mu_L =$ liquid viscosity, cp.

N_0 is the minimum necessary agitation speed for the uniform dispersion of bubbles in a stirred tank and N_0 was expressed as follows from their experimental results:

$$N_0 D/(\sigma g/\rho_L)^{0.25} = 1.22 + 1.25(T/D) \quad (69)$$

where N_0 = minimum necessary agitation speed, R.P.H.
σ = surface tension, kg/hr^2.

Miller (1974) recently developed procedures for predicting liquid-film-controlled mass transfer in gas-sparged contactors with an dwithout mechanical agitation. Experimental measurements were made on CO_2 stripping from aqueous solution with air in 0.00252, 0.0252, and 0.252 m^3 tanks which were geometrically similar, fully baffled, and equipped with four flat-blade impellers and spargers. Mass transfer is shown to depend on mean bubble size. Bubble shape, motion, and interface fluctuation are all properties that are associated with bubble size, which, in turn, can be determined from the physical characteristics of the contacting system.

Recently, the gas absorption by non-Newtonian fluids was studied by Perez and Sandall (1974), investigating the absorption of carbon dioxide into aqueous Carbopol solutions in a turbine-agitated vessel for the cases of absorption across an unbroken interface and absorption with the gas bubbling through the liquid. For gas absorption across an unbroken interface, the data are correlated with a standard deviation of 8.5% by

$$\frac{k_c d}{D} = 5.11 \times 10^{-3} \left(\frac{d^2 N \rho}{\mu_e}\right)^{0.926} \left(\frac{\mu_e}{\rho D}\right)^{0.5} \tag{70}$$

For absorption in a mechanically agitated gas/liquid dispersion in the absence of surface foam, the data are correlated with a standard deviation of 5.0% by

$$\frac{k_c a d^2}{D} = 21.2 \left(\frac{d^2 N \rho}{\mu_e}\right)^{1.11} \left(\frac{\mu_e}{\rho D}\right)^{0.5} \left(\frac{d V_s}{\sigma}\right)^{0.447} \left(\frac{\mu_g}{\mu_e}\right)^{0.694} \tag{71}$$

The effective viscosity for the liquids was obtained by

$$\mu_e = K(11N)^{n-1} \left(\frac{3n+1}{4n}\right) \tag{72}$$

where a = effective gas/liquid interfacial area per unit volume, cm^{-1}
D = diffusion coefficient, cm^2/sec
d = impeller diameter, cm
k_c = mass transfer coefficient, cm/sec
K = consistency index, $g/cm.sec^{2-n}$
N = agitation speed, R.P.S.
n = flow behavoir index
V_s = superficial gas velocity through sparge tube, cm/sec
μ_e = effective liquid viscosity, $g/cm \cdot sec$
μ_g = gas viscosity, $g/cm \cdot sec$
ρ = liquid density, g/cm^3
σ = surface tension, dynes/cm

6. Concluding Remarks

In the foregoing pages, oxygen transfer and scale-up in submerged aerobic fermentation have been discussed on the basis of author's and other investigations. There are still many unsolved problems in this field. For example, mass transfer and agitation power in highly viscous or non-Newtonian culture fluids have not been sufficiently investigated. The non-uniform distributions of dissolved oxygen concentration, gas bubble size, oxygen partial pressure in the bubbles, residence time, and mass-transfer coefficient are often observed in highly viscous or non-Newtonian culture media. The oxygen-transfer rates in these fermentations should be obtained taking account of their non-uniform distributions as proposed by Taguchi *et al.* (1972). The effect of shear is an important problem in the scale-up of aerated agitated fermenters as shown by a few investigators (Midler and Finn, 1966; Taguchi and Yoshida 1968) and it has to be further investigated. Interesting investigations on the mechanism of oxygen transfer in submerged aerobic fermentation were reported by Tsao (1968, 1969, 1970) and Wise *et al.* (1969) but fundamental studies on the mechanism still remain to be performed.

Nomenclature

a = gas/liquid interfacial area per unit liquid volume, m^2/m^3
a_m = liquid/cell interfacial area per unit liquid volume, m^2/m^3
a_p = liquid/pellet interfacial area per unit liquid volume, m^2/m^3
B = width of baffle, m
C = concentration of dissolved oxygen or penetrating component, kg-moles/m^3
$(C_L)_{cr}$ = critical oxygen concentration in liquid, kg-moles/m^3
$(C_{Li})_{cr}$ = critical oxygen concentration at liquid/cell or liquid/pellet interface, kg-moles/m^3
C_L = concentration of dissolved oxygen in bulk of liquid, kg-moles/m^3
$(C_L)_i$ = initial concentration of dissolved oxygen in bulk of liquid, kg-moles/m^3
C_{Li} = concentration of dissolved oxygen at liquid/cell or liquid/pellet interface, kg-moles/m^3
C_L^* = oxygen concentration in equilibrium with air, kg-moles/m^3
C_0 = concentration of dissolved oxygen at center of cell or pellet kg-moles/m^3
C_m = dry cell weight per unit liquid volume, kg/m^3
D = agitator diameter, m
D_e = effective diffusion coefficient of dissolved oxygen in mycelial pellet, m^2/hr
D_L = molecular diffusion coefficient of dissolved oxygen or transfer material in liquid, m^2/hr
D_p = molecular diffusion coefficient of dissolved oxygen in mycelial pellet, m^2/hr
d_{BM} = average diameter of bubble, m
d_p = diameter of particle, m
g = acceleration due to gravity, m/hr^2
g_c = gravitational conversion factor, kg · m/Kg · hr^2, where Kg indicates gravitational unit, while kg indicates mass unit
H = depth of medium in tank, m
H_G = holdup of gas in tank
H_L = holdup of liquid in tank
h = clearance from tank bottom to agitator, m
k_c = specific respiration coefficient defined by Eq. (15)
K_L = overall mass-transfer coefficient based on liquid film around gas bubbles, m/hr
k_L = mass-transfer coefficient for liquid film around gas bubbles, m/hr
K_m = apparent Michaelis constant for mycelia, kg-moles/m^3
K'_m = apparent Michaelis constant for pellet, kg-moles/m^3
k_m = mass-transfer coefficient for liquid film around cells or pellets defined by Eq. (2), m/hr

k_p = oxygen-transfer coefficient for pellet defined by Eq. (27), m/hr

k_r = specific oxygen-uptake rate per unit dry mycelial weight, kg-moles of oxygen/[(hr)(kg of dry cell)]

$(k_r)_{max}$ = maximum specific oxygen-uptake rate per unit dry mycelial weight, kg-moles of oxygen/[(hr)(kg of dry cell)]

l = length of agitator blade, m

N = agitation speed, R.P.H. or R.P.M.

N_{Fr} = Froud number defined by DN^2/g

N_p = Power number defined by $P_0 g_c/\rho_L N^3 D^5$

N_{Pe} = Peclet number defined by $d_P U/D_L$

N_{Re} = Reynolds number defined by $D^2 N \rho_L/\mu_L$

$(N_{Re})_P$ = Reynolds number for pellet defined by $d_P D N \rho_L/\mu_L$

P_g = power consumption accompanied by aeration, Kg · m/hr

P_0 = power consumption without aeration, Kg · m/hr

P_{max} = maximum power consumption without aeration, Kg · m/hr, where Kg indicates gravitational unit

q = volumetric flow rate of gas, m^3/hr

R = radius of mycelial pellet, m

r = radial distance from center of cell or mycelial pellet, m

r_L = radial distance from cell center to liquid bulk, m

r_0 = radius of cell, m

T = diameter of tank, m

$t_{1/2}$ = half-life time of oxygen concentration in liquid, min

U = relative velocity of particle and fluid, m/hr

U_t = terminal velocity of particle in free fall, m/hr

V = liquid volume in tank, m^3

V_r = rapid oxygen-uptake rate per unit mycelial pellet, given by Eq. (21), kg-moles of oxygen/[(hr)(one pellet)]

V_r' = oxygen-uptake rate per unit mycelial pellet, kg-moles of oxygen/[(hr)(one pellet)]

V_s = superficial gas velocity based on cross section of tank, m/hr

V_t = terminal gas velocity in free rise, m/hr

W = agitator blade width, m

x = distance for penetration direction from interface, m

ϵ = void fraction

η = effectiveness factor for oxygen consumption rate per unit mycelial pellet

θ = time, hr

μ_L = liquid viscosity, kg/m · hr

ρ = mycelial density in pellet, kg/m^3

ρ_L = liquid density, kg/m^3

$\Delta\rho$ = difference in density between dispersed and continuous phases, kg/m^3

References

Aiba, S.: Shin-kagakukogakukoza, Vol. 24, "Fermentation Technology", The Industrial Daily News, Japan (1960).

Aiba, S., Kobayashi, K.: Biotechnol. Bioeng 13, 583 (1971).

Bartholomew, W. H.: Advances in Applied Microbiology **2**, 289 (1960).

Bartholomew, W. H., Karow, E. O., Sfat, M. R., Wilhelm, R. H.: Ind. Eng. Chem. **42**, 1801 (1950).

Blakebrough, N., Sambamurthy, K.: Biotechnol. Bioeng 8, 25 (1966).
Brian, P. L. T., Hales, H. B.: AIChE Journal **15**, 419 (1969).
Bylinkina, E. S., Birukov, V. V.: Proceedings of the 4th International Fermentation Symposium, Page 105, Kyoto (1972).
Calderbank, P. H.: Trans, Inst. Chem. Engrs. **36**, 443 (1958).
Calderbank, P. H.: Trans. Inst. Chem. Engrs. **37**, 173 (1959b).
Calderbank, P. H., Moo-Young, M. B.: Trans. Inst. Chem. Engrs. **37**, 26 (1959a).
Calderbank, P. H., Moo-Young, M. B.: Chem. Eng. Sci. **16**, 39 (1961).
Cooper, C. M., Fernstrom, G. A., Miller, S. A.: Ind. Eng. Chem. **36**, 504 (1944).
Friedman, A. M., Lightfoot, E. N., Jr.: Ind. Eng. Chem. **49**, 1227 (1957).
Fukuda, H., Sumino, Y., Kanzaki, T.: J. Ferment. Technol. **46**, 829 (1968).
Gaden, E. L. Jr.: Sci. Repts. Instituto Superiore di Sanita **1**, 161 (1961).
Harriott, P.: AIChE Journal **8**, 93 (1962).
Hixson, A. W., Gaden, E. L. Jr.: Ind. Eng. Chem. **42**, 1792 (1950).
Huang, M. Y., Bungay, H. R. 3rd: Biotechnol. Bioeng **15**, 1193 (1973).
Humphrey, A. E.: J. Ferment. Technol. **42**, 265 (1964a).
Humphrey, A. E.: J. Ferment. Technol. **42**, 334 (1964b).
Kalinske, A. A.: Sewage and Ind. Waste, **27**, 572 (1955).
Kobayashi, J., Ueyama, H.: J. Ferment. Technol. **40**, 63 (1962).
Kobayashi, T., Dedem, G. V., Moo-Young, M. B.: Biotechnol. Bioeng **15**, 27 (1973).
Levich, V. G.: Physicochemical Hydrodynamics, Prentice-Hall, Englewood Cliffs, New Jersey, 1962.
Maxon, W. D.: J. Biochem. Microbiol. Techn. Eng. **1**, 311 (1959).
Michel, B. J., Miller, S. A.: AIChE Journal **8**, 262 (1962).
Midler, M., Jr., Finn, R. K.: Biotechnol. Bioeng **8**, 71 (1966).
Miller, D. N.: AIChE Journal **20**, 445 (1974).
Miura, Y.: Dissertation, Kyoto University, Kyoto, Japan, 1961.
Miura, Y., Hirota, S.: J. Ferment. Technol. **44**, 890 (1966).
Miura, Y., Kanamori, T., Miyamoto, K.: 1st Pacific Chemical Engineering Congress, Kyoto, Japan (1972).
Miura, Y., Miyamoto, K., Kanamori, T., Teramoto, M., Ohira, N.: in press (1975)
Miura, Y., Miyamoto, K., Matsuda, R.: Preprint of 88th Annual meeting of The Pharmaceutical Society of Japan, Tokyo, **443** (1968).
Nagata, S., Yamaguchi, I., Nishikawa, M., Wada, K.: Chem. Eng. (Japan), **31**, 1016 (1967).
Nagata, S., Yamamoto, K., Yokoyama, T., Shiga, S.: Chem. Eng. (Japan) **21**, 708 (1957).
Nagata, S., Yokoyama, T., Maeda, H.: Chem. Eng. (Japan) **20**, 582 (1956).
Oyama, Y., Endoh, K.: Chem. Eng. (Japan), **19**, 2 (1955).
Perez, J. F., Sandall, O. C.: AIChE Journal **20**, 770 (1974).
Phillips, D. H.: Biotechnol. Bioeng 8, 456 (1966).
Richards, J. W.: Progr. Ind. Microbiol. **3**, 141 (1961)
Robinson, C. W., Wilke, C. R.: Proceedings of the 4th International Fermentation Symposium, 73 (1972).
Roxburgh, J. M., Spencer, J. F. T., Sallans, H. R.: Agr. Food Chem. **2**, 1121 (1954).
Rushton, J. H., Costich, E. W., Everett, J. H.: Chem. Eng. Progr. **46**, 395 (1950).
Satterfield, C. N.: Mass Transfer in Heterogeneous Catalysis, M. I. T. Press (1970).
Steel, R., Maxon, W. D.: Biotechnol. Bioeng **4**, 231 (1962).
Steel, R., Maxon, W. D.: Biotechnol. Bioeng 8, 97 (1966a).
Steel, R., Maxon, W. D.: Biotechnol. Bioeng **8**, 109 (1966b).
Strohm, J. A., Dale, R. E., Peppler, H. J.: Appl. Microbiol. 7, 235 (1959).
Taguchi, H., Miyamoto, S.: Biotechnol. Bioeng 8, 43 (1966).
Taguchi, H., Suga, K., Yoshida, T.: Proceedings of the 4th International Fermentation Symposium, 83 (1972).
Taguchi, H., Yoshida, T.: J. Ferment. Technol. **46**, 814 (1968).
Terui, G., Konno, N., Sase, M.: Techn. Report of Osaka Univ. **10**, 527 (1960a).
Terui, G., Konno, N., Sase, M.: J. Ferment. Technol. **38**, 278 (1960b).

Tsao, G. T.: Biotechnol. Bioeng **10**, 765 (1968).
Tsao, G. T.: ibid. **11**, 1071 (1969).
Tsao, G. T.: ibid. **12**, 51 (1970).
Tsao, G. T., Mukerjee, A., Lee, Y. Y.: Proceedings of the 4th International Fermentaion Symposium, 65 (1972).
Vermeulen, T., Williams, G. M., Langlois, G. E.: Chem. Eng. Progr., **51**, 85F (1955).
Wegrich, O. G., Shurter, R. A.: Ind. Eng. Chem. **45**, 1153 (1953).
Westerterp, K. R., Dierendonck, L. L. van, Dekraa, J. A.: Chem. Eng. Sci. **18**, 157 (1963).
Wise, D. L., Wang, D. I. C., Mateles, R. I.: Biotechnol. Bioeng. **11**, 647 (1969).
Yano, T., Kodama, K., Yamada, K.: Agr. Biol. Chem. **25**, 580 (1961).
Yano, T., Yamada, K.: Institute Appl. Microbiol. Symp. on Microbiol., No. 5, 16 (1963).
Yoshida, F., Ikeda, A., Imakawa, S., Miura, Y.: Ind. Eng. Chem., **52**, 435 (1960).
Yoshida, F., Miura, Y.: Ind. Eng. Chem. Process Design and Development **2**, 263 (1963).
Yoshida, T., Shimizu, T., Taguchi, H., Teramoto, S.: J. Ferment. Technol. **45**, 1119 (1967).
Yoshida, T., Taguchi, H., Teramoto, S.: J. Ferment. Technol. **46**, 119 (1968).

CHAPTER 2

Microbial Flocs and Flocculation in Fermentation Process Engineering

B. ATKINSON,
UMIST, Dept. of Chemical Engineering, University of Manchester, Institute of Science and Technology, P. O. Box 88, Manchester, M 60 1 QD, Great Britain

I. S. DAOUD,
Dept. of Chemical Engineering, Engineering College, Baghdad University, Baghdad, Iraq

With 43 Figures

Contents

1. Introduction

Freely suspended microorganisms often occur in groupings or flocs with a characteristic size many orders of magnitude greater than that of a single microorganism. The factors producing a particular floc size are largely unknown but flocculation can be enhanced by the presence of so-called flocculating agents, e.g. aluminium and calcium chloride (Aiba and Nagatani, 1971), or floc size can be decreased by increased shear (Aiba and Nagatani, 1971) or surface-active agents (Mallette, 1969). At the present time it seems reasonable to suggest that all organisms can occur either as single cells or in multi-cell groups; however, some microorganisms have a tendency to form larger flocs than others, under what are apparently the same environmental conditions. In fact, flocculation characteristics are sometimes used for purposes of classification, e.g. yeasts are listed on a relative basis as flocculating or non-flocculating (Hough, 1957).

The productivity of growth-associated fermentations is dependent upon the quantity of biomass contained in the fermenter (microbial hold-up). In contrast the performance characteristics of a fermenter are dominated by the form in which the microbial hold-up occurs, i.e. whether in suspension, or adhering to a support surface, or a combination of both. For fermenters containing organisms in suspension the agglomeration of single suspended organisms into discrete particles or flocs is a significant factor in the functioning of the overall process, i.e. fermenter and biomass separation stage. The performance characteristics, the mass and heat transfer characteristics, the power requirements and the productivity of a fermenter may be affected by the presence of flocs owing to:

a) increased microbial hold-up;
b) ease of separation of biomass from the fermentation broth;
c) possibilities for the use of "novel" fermenter configurations;
d) improved rheology at high biomass concentrations.

Against these has to be set the fact that even if the environmental conditions within a floc are the same as the external environment, then the overall specific substrate uptake is likely to be reduced because of diffusional limitations on the transfer of substrate through the floc. However, ZoBell (1943) has pointed out that there is a possible advantage to the cells in the retardation of metabolites and exo-enzymes by diffusion, as this may result in the production of the optimum physico-chemical conditions within the biomass.

A summary of the processes in which the occurrence of flocs has been found advantageous is given in Table 1.

Table 1. Summary of processes using flocs

Process objectives and culture (References)	Research objectives and factors investigated	Fermentation methods and conditions	Characteristics of flocs	Control of floc size	Measurement of floc size and flocculation
Biomass production using *A. niger* (Morris *et al.*, 1973; Smith and Greenshields, 1974)	Use of tower fermenters for biomass production. Aeration, physicochemical conditions	Medium: supplemented molasses; air velocity: 1.0–6.0 cm sec^{-1}; pH: 3.25–6; time: upto 48 hrs; biomass concentration: 0.25–1.125% w/w; tower fermenters (1, 10, 50, 1000 l, aspect ratio 10 : 1): batch and continuous operation	Ranged from filamentous threads through degrees of hairy pellets to smooth rounded colonies; up to 35 mm dia		Photographic and direct measurement
Biomass production using *S. cerevisiae*, NCYC 1250 and 1251 (Smith and Greenshields, 1974)	Use of tower fermenters for biomass production. Yeast strain, media composition, gas and liquid flowrates	Medium: malt extract based wort (SG 1.03 and 1.014); air flow-rates: 2 and 7 l min^{-1}; media flow-rates: 12 and 15 l day^{-1}; max. yeast concs: 3.1% w/w (NCYC 1250), 17.5% w/w (NCYC 1251); tower fermenters (5 l, 6 l, aspect ratio 10 : 1) batch and continuous operation	5 l tower, flocs (0.1–1.0 cm dia. NCYC 1251) of various shapes, many disc-like; 6 l tower, small flocs (< 0.1 cm dia, NCYC 1250) readily washed out	Strain selection	Photogaphic and direct measurement
Brewing using *S. cerevisiae* (Greenshields and Smith, 1971)	Use of tower fermenter for brewing. Yeast strain, wort density and flow rate	Medium: wort (S. G. 1.040): space velocity: 0.3–0.2 day^{-1}; mean yeast conc. 25% w/w centrifuged wet weight; tower fermenter, continuous operation	a) physically limited, attains yeast conc. of 20–30% w/w: upto 0.2 cm dia; b) fermentation limited, attains yeast conc. of 25–40% w/w: upto 1.3 cm dia	Strain selection, wort specific gravitiy, superficial liquid velocity	Photographic and direct measurement

Table 1. (continued)

Process objectives and culture (References)	Research objectives and factors investigated	Fermentation methods and conditions	Characteristics of flocs	Control of floc size	Measurement of floc size and flocculation
Citric acid production using *A. niger* (Steel *et al.*, 1955)	Scale-up and factors affecting citric acid production in submerged culture. Ferrocyanide treatment and phosphate supplement of molasses; inoculum level; pH; temperature and aeration rate	Inoculum: $120-280 \times 10^3$ pellets l^{-1}; aeration rate: a) 500 ml. min^{-1} b) 6400 ml. min^{-1}; initial pH: 5.0–7.0; temp: 31° C; time: 116 hrs; tower fermenters a) 2.5 l and b) 36–40 l; batch operation	Varied in shape and density; optimum for citric acid production, 1–2 mm dia	Size of spore inoculum, pH and ferrocyanide level	Photographic and direct measurement
Enhancement of cell recovery in microbial protein production using the bacteria *P. fluorescens, E. coli* and *L. delbrueckii* (McGregor and Finn, 1969)	Use of chemical flocculents. Bacterial genus, chemical nature and concentration of a cationic polyamine, and small, positively charged, alumina fibrils	pH: neutral; flocculation test: 2 l suspension agitated on a shaker for 20 min; flocculation vessel: 2.5 l; batch		Type and concentration of flocculants	Amount of flocculant required to achieve 50% removal of cells after 1 min of centrifugation at 2250 r.p.m.
Enhancement of cell recovery in microbial protein production using *Candida intermedia* (Gasner and Wang, 1970)	Use and evaluation of 50 chemical flocculants. Chemical nature and concentration of additive; effect of shear	Air flow: 1.0 min^{-1}; pH: 3; temp: 30° C; yeast conc: 4.0 gm/l; flocculation vessel: 500 ml; flocculation test: 150 ml broth agitated on a shaker for 5 min; fluid mechanics test: annular flow viscometer; stirred fermenter (14 l): semi-continuous operation		Type and concentration of additive; shear	Volume of sediment and residual turbidity after 20 min settling

Table 1. (continued)

Process objectives and culture (References)	Research objectives and factors investigated	Fermentation methods and conditions	Characteristics of flocs	Control of floc size	Measurement of floc size and flocculation
Mushroom Mycelium production using *Agaricus blazei* (Block *et al.*, 1953)	Cultivation of mushroom on citrus press water in submerged culture. pH, size of fermenter and media	pH: 5.5–8.0; 250 ml flask, 1 l and 4 l bottles; batch operation; flask agitated on shaker; bottles stirred and aerated	Shake flasks: 3.2 mm dia. balls; aerated bottles: 2.54 mm dia. clumps, but much less dense than the smaller clumps	Type of agitation, aeration, and sugar concentration of medium	Microscopic
Plant tissue culture using carrot (Smith and Street, 1974)	Decline in embryogenic potential in carrot culture. Subculturing every 21 days	pH: 5.5; temp: 25° C; biomass concentration: 6.4–10.0 g l^{-1}; continuous light; 100 ml shake flask: batch operation	Tissue grew as embryogenic clumps 100–500 μm in dia		
Plant tissue culture using endosperm of corn (Graebe and Novelli, 1966)	Method for large scale production. Time of growth and procedure of cultivation	Medium: synthetic based on sucrose; pH: 6.0; temp: 25–28° C; time: 71 days; tissue concentration: 24–168 g l^{-1}; 6 l flasks: semicontinuous operation. Flasks aerated and stirred intermittently	Tissue grew as aggregates few mm dia. when intermittent stirring was used, otherwise grew up to 3 cm dia	By intermittent stirring	
Tissue culture of embryonic cells. Liver, retina, kidney, and limb–bud from chick and mouse embryos (Moscona, 1961a)	Factors affecting aggregation of trypsin dissociated cells. Type of organism and cell, age of embryo, temperature, rotation speed, presence of inhibitors or promoters	pH: 8.0; temp: 15–38° C; time: 24 hrs; 25 ml shake flask: batch operation	up to 1 mm dia	Age of embryo, temperature and shaker r.p.m.	Photographic

Table 1. (continued)

Process objectives and culture (References)	Research objectives and factors investigated	Fermentation methods and conditions	Characteristics of flocs	Control of floc size	Measurement of floc size and flocculation
Waste water treatment using activated sludge (Pavoni *et al.*, 1972)	Mechanism of flocculation. Physiological state of microorganisms, accumulation of exocellular polymers, pH, agitation	Initial COD: 2000 mg/l; pH: 1–12; time: up to 140 hrs; biomass conc: 1000–1700 mg/l; batch (15 l): aeration and agitation by air diffusers			Filtration time; light transmission and turbidity following settling
Waste water treatment using activated sludge (DeWalle and Chian, 1974)	Formation and effect of humic substances on flocculation	Inoculum: 20100 mg l^{-1}; pH: 0.2–8; time: up to 53 hrs; suspended solids: 140–260 mg l^{-1}; d.o.: 1.3–8 mg l^{-1}; initial total organic carbon 91 mg l^{-1}; batch (80 l): aeration rate 28 l min^{-1}		When carbohydrate content of humic substances was low adsorption onto cells decreased with impairment of floc formation and settleability	Turbidity after filtration through Whatman No. 4 filter paper

1.1 Rate and Design Equations

In chemical reactor theory the empirical nth order rate equation for reaction kinetics has resulted in the development of equations which describe the operating characteristics of "ideal" reactors. These design equations have proved invaluable as an aid to the understanding of the performance of "real" reactors, in the development of design procedures for "real" reactors, and in devising reactor configurations for particular duties. In laboratory studies the concept of the ideal reactor has assisted in the design of experiments, in the development of experimental apparatuses, and in the interpretation of experimental data. Thus it is reasonable to anticipate that in biochemical reactor theory the availability of a generalised rate equation for the overall rate of substrate uptake by microbial flocs is a necessary prerequisite for the development of generalised design procedures, and for the interpretation of experimental data (Atkinson, 1974).

1.2 Fermenter Configurations

The conventional fermenters containing organisms in suspension are the batch fermenter (BF) and the continuous stirred tank fermenter (CSTF). If the particles are contained in a tube without a stirrer and the feed enters at the base and the product leaves at the top, an entirely new situation arises. At low flow-rates the particles are stationary and the flow passes through the bed by a tortuous path; this is the so-called "fixed bed" arrangement. As the flow-rate is increased the pressure loss in the fluid as it passes through the bed becomes equal to the weight of the bed, and in consequence the particles become suspended in the fluid largely out of contact with one another. The bed of particles still retains its identity and a well-defined bed length exists, albeit greater in magnitude than the original 'fixed' bed length. Further increases in velocity ultimately result in the particles being swept from the tube. The first and last of these fluid-solid flow patterns can be described as tubular fermenter configurations, i.e. the substrate concentration falls and the product concentration rises in the direction of flow. The intermediate situation represents an example of a 'fluidised' bed.

Reactors based upon each of the above patterns are a feature of the chemical industry. The range of allowable flow-rates for each condition is normally quite large; this is a consequence of the particle sizes used and the differences in density between fluid and solid (Coulson and Richardson, 1968).

In contrast to inert particles, biological particles 'grow'; they can have a wide size range and are of a density very similar to that of the fluid phase. The first of these characteristics largely rules out the possibility of the fixed-bed operation of a fermenter containing flocs, because the particles in contact tend to 'grow' together. This results in a most irregular flow path for the liquid, and in particles of an excessively large characteristic size (see Section 5).

It follows that a tubular fermenter configuration involving microbial flocs is feasible only when the particles are carried along with the fluid. Such an arrangement demands a constant supply of microorganisms to the base of the fermenter. This can be achieved on a recycle basis, though it must be appreciated that the physiological and biochemical characteristics of the microorganisms returned to the fermenter will reflect the environmental conditions of the outlet rather than the inlet.

The processes taking place in a tubular fermenter are significantly more complex than those in the CSTF, since the individual microbial flocs are exposed to differing environmental conditions during their passage through the fermenter. The tubular fermenter, when operated at a high recirculation rate, is equivalent to the CSTF and the flow pattern is then similar to that which exists in the air-lift fermenter (Wang and Humphrey, 1969).

It is hardly surprising, in view of the complexity of an apparently simple situation, that the tubular fermenter containing microbial flocs has yet to establish itself in the fermentation industries.

1.3 Terminology

The occurence of bacteria, yeasts, fungi, plant and animal cells, as multicellular colonies, has led to the use of a varied and imprecise terminology. The reasons for this are largely to be found in the scientific and technological divisions under which the various organisms have been studied. Virtually the same phenomena has been described as aggregation, clumping, flocculation, and pellet formation. The term aggregation has most often been used with embryonic cells, while bacterial cultures are described as clumping. On the other hand the term flocculation is most commonly associated with activated sludge and brewers' yeast, whilst pellet formation is the most common term with fungal organisms. It can be argued that the use of a particular term has implications regarding the mechanism of formation of a particular multicellular colony; however, these mechanisms are not particularly clear-cut or mutually exclusive and there seems little justification for the use of the various terms except interchangably.

2. Aggregation Mechanisms

2.1 Polymer-Colloid Flocculation

Study of the mechanism of aggregation of colloids when using polymers is important in water treatment and in soil conditioning, and should be of assistance in understanding the role of polymers in microbial aggregation. Colloids may be hydrophilic or hydrophobic in nature, while water-soluble polymers can be either charged or uncharged and hence classified as polyelectrolytes or non-ionic polymers respectively.

Walles (1968) classified the various flocculation mechanisms as:

charge effect neutralization;

polymer bridging; and

mutual dehydration.

These mechanisms are not mutually exclusive. In charge neutralization the net effect of the electrostatic repulsion between particles having an electric double layer is neutralised. Added polyelectrolytes compress the electric double layers permitting Van der Waal's forces to draw the particles towards each other and hold them together. The loss of all or part of the electrical double layer by chemical reaction with a polyelectrolyte is also included. In polymer bridging soluble polymer is adsorbed on the solid surfaces, each polymer chain adsorbing on, and bridging between, more than one solid particle. The polymer attachment is the result of specific chemical interactions with the surface, e.g. the formation of chemical complexes, hydrogen bonding, and proton transfer. A particle with polymeric chains extending into the bulk of the solution, is considered to behave in collisions as if it had an increased radius. Interactions occur between the polymeric chains on adjacent particles and this results in bridging between the dispersed particles and the growth of aggregates (Michaels, 1954). In mutual dehydration a chemical reaction between an hydrophilic colloid and a non-ionic polymer results in an insoluble non-colloidal particle. Such particles are readily separable from solution, but no aggregation takes place.

The stability of suspensions of hydrophobic colloids, mainly inorganics, is largely due to the surface potential. It follows that the interaction between such colloids and polyelectrolytes may lead to destabilization as a result of charge neutralization. In contrast most hydrophilic colloids, mainly organics, receive their charge and stability from the ionization and hydration of certain functional groups on the surface. Thus aggregation is most likely to result from polymer bridging. Consequently the hydrophilic nature of biocolloids, and the similarity in charge between biopolymers and biocolloids, generally favours polymer bridging as the mechanism of formation of biological aggregates.

Factors affecting interactions between colloidal surfaces and polymers include: particle size and density, the concentration, configuration and molecular weight of the polymer, pH and ionic properties of the solution, and the intensity and time of agitation.

In the polymer-bridging mechanism, it is recognized that the optimal polymer concentration for aggregation will be a function of the total available surface area of the colloidal suspension. Linke and Booth (1960) observed a linear relationship for this dependency and suggested that, when aggregates were broken apart at low aggregate concentrations, the time of free travel was so large that the loose ends of the adsorbed polymer molecules (former bridges) had a greater chance of becoming adsorbed to the same colloidal particle, thus reducing the total amount of polymer bridging in the system. The concentration of solids affects the optimal polymer/solid ratio as a result of the differential rates of polymer adsorption and inter-particular collisions. As a result of infrequent collisions and polymer rearrangement on the surface of the solid, low solids concentrations lead to less aggregation; as the concentration of solids increases, more aggregates form.

Studying the flocculation of silica with cationic polymers in the concentration range 0–10 p.p.m., Dixon *et al.* (1967) have shown (Fig. 1) that polymers with a molecular weight up to 5 200 exhibit little effect on filtrability, and therefore on flocculation, at concentrations beyond the optimum. In contrast higher molecular weight polymers caused redispersal of the flocs and poor filtrability at polymer concentrations beyond the optimum. Similar behaviour was noted by Healy (1961) for quartz flocculated with the aid of partially hydrolyzed polyacrylamide; a cationic flocculant.

The above-mentioned systems were concerned with the flocculation of negatively charged solid particles using cationic polyelectrolytes. Michaels (1954) has shown that, for aggregation of soils using anionic polyelectrolytes, e.g. hydrolysed polyacrylamide, the degree of flocculation increases rapidly with increasing polymer concentration up to a value of

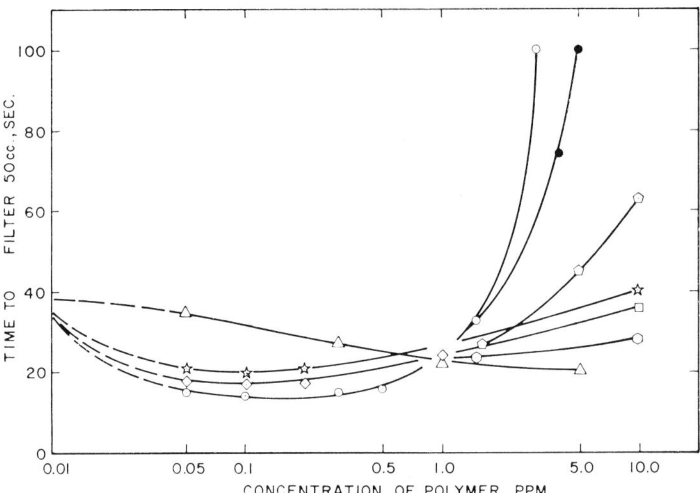

Fig. 1. Effect of molecular weight of cationic polymer on filtration of 5 μm silica (pH 6; 0.04 wt % silica). Molecular weights: ○ 32000; ◇ 5200; □ 1700; ☆ 1000; △ 189; ●, ⌂, polymers cross-linked with epichlorohydrin (Dixon *et al.*, 1967)

about 0.03 gm per 100 gm of dry soil. Above this concentration the extent of flocculation is increased only slightly, if at all, and in some cases actually decreased. Reporting on the influence of the molecular weight of linear long-chain water-soluble polymers on the flocculation rate, Walles (1968) presents experimental data, obtained by Smith and Volk (1968) for clay-polysodium styrene sulfonate (SPSS), and by Manary (1968) for bauxite red mud liquor-SPSS, which clearly demonstrate the sensitivity of the flocculation process to the molecular weight of the flocculant, the degree of flocculation being higher for the higher molecular weight species. The same conclusion was reached by Michaels (1954). It was concluded by Dixon *et al.* (1967) that as the molecular weight of tetraethylene pentamine (TEPA) increased from 189 to 1000 the concentration required for good flocculation decreased some 100-fold. Above 1000 and to 32000 the effect of molecular weight was small. When the polymer molecules are linear and rigid, and exist in solution as fairly stiff extended rods, then there is a direct correspondence between molecular weight and chain length. Michaels (1954) puts forward the idea that some polymers have flexible chains which can fold, kink, or coil in solution owing to intramolecular bonding. Such a phenomena would inactivate groups which have a latent adsorptive affinity for solid surfaces and greatly reduce the effective length of the molecule in solution. A polymer molecule may be "stretched out" by introducing ionized groups at intervals into its chain. The type and quantity of the ionized groups would necessarily depend on the system involved.

The pH and ionic characteristics of the solution have been shown to have a profound effect on flocculation. These characteristics include the surface potential (zeta potential), the charge and nature of the double layer surrounding each particle, the charge and charge density of the polymer. Dixon *et al.* (1967) showed that the electrophoretic mobility of silica is sensitive to the presence of low concentrations of cationic polymer of any molecular weight (Fig. 2). In addition the charge on the silica changed from negative to positive in the polymer concentration range which coincided with the range where the best flocculation was attained. Further increase in polymer concentration caused mobility to increase rapidly but had little effect on flocculation. At even higher concentrations mobility remained the same but eventually a redispersal of the silica occured.

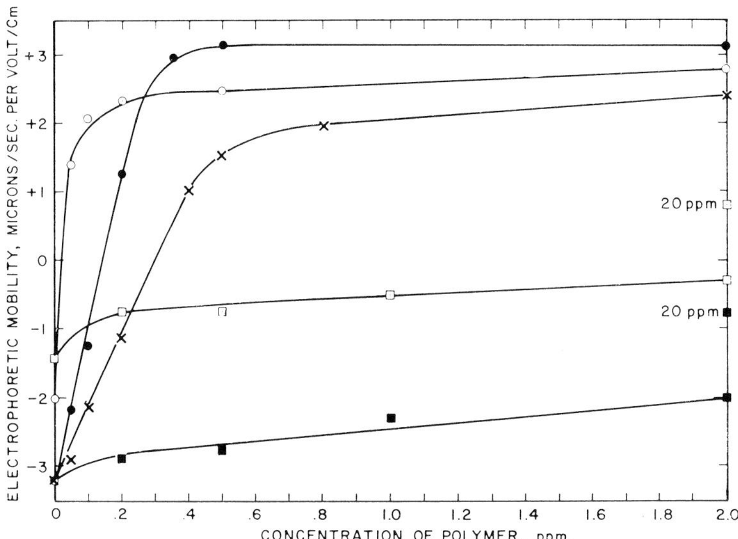

Fig. 2. Effect of polymer molecular weights on the electrophoretic mobility of 5 μm silica (0.04 wt %). Molecular weights and pH: ■ 189, 9; □ 189, 4; X 1000, 9; 0 1700 and 3200, 4; ● 1700 and 3200, 9 (Dixon *et al.*, 1967)

Dixon *et al.* (1967) also show that a decrease in pH in the range 9.0 to 4.0 lowered the diethylene triamine (DETA) polymer concentration needed for maximum flocculation and zero mobility (Fig. 2) and presumed this to be due to an increase in protonation of the amino groups as the pH was lowered.

Healy (1961) reported a rapid decrease in the zeta potential of quartz with increasing polymer concentration, but the potential was not reduced to zero (Table 2). After adsorption the quartz was washed with water and the zeta potential redetermined; no difference was found below a polymer concentration of 0.1 g/litre (Table 2). The adsorption of polymer was therefore considered to be irreversible to washing at concentrations up to 0.1 g/litre and partially reversible above this value.

Regarding anionic polymers Michaels (1954) claims that a low zeta potential solid can adsorb a highly anionic polymer, and vice versa, but that there is a maximum negative zeta potential above which flocculation with anionic polyelectrolytes is impossible. Examples of the latter are aluminosilicate soil and clay suspensions, at high pH in the presence of alkali ions (zeta potential is large), which do not flocculate with anionic polyelectrolytes. Anionic polyelectrolytes are active as flocculants only in soils and clays suspended in the presence of free electrolyte; for under these conditions the zeta

Table 2. Effect of water washing on zeta potential of quartz previously exposed to separan 2610 (Healy, 1961)

Concentration of Separan 2610 (g/litre)	Zeta potential of quartz in test solution (mv.)	Zeta potential[a] of quartz after washing (mv.)
0	−63.0	−63.0
0.001	−43.5	−44.0
0.010	−31.5	−31.5
0.050	−21.5	−22.0
0.100	−17.5	−18.0
0.175	−16.0	−16.5
0.250	−15.5	−16.5
0.500	−13.5	−16.5
0.750	−12.5	−16.5

[a] Determined in conductivity water.

potentials are low. Michaels also suggests that the degree of ionization (and hence the polymer charge density) can be adjusted by the use of carboxylate or other weakly acidic groups as chain "extenders", and can be relatively easily controlled by changing the pH, or adding polyvalent metal ions, e.g. calcium, which can form chelates with these groups.

Discussing results obtained on the effect of time and intensity of agitation on flocculation, Healy (1961) states that: with increasing intensity of agitation the amount of polymer adsorbed and the average floc size decrease; with increasing time of agitation at a fixed intensity, the amount of polymer adsorbed does not change but the average floc size decreases.

2.2 Microbial Flocculation

Microbial flocculation is currently of particular importance in many biochemical processes, e.g. in brewing, in the activated sludge process, and in biomass production. In discussing the mechanisms of flocculation, classification based on the type of microorganism involved, rather than the process, is more satisfactory since certain types are common to more than one process. Thus the flocculation of yeasts, bacteria, moulds, and mixed populations can be discussed independently of the fermenter or process configuration and objectives. In addition the culturing of plant cells and tissue culture (embryonic cells) involves flocculation phenomena.

The mechanism of aggregation of cells has not been established fully as yet. Knowledge is sparse in the case of bacterial aggregation, while yeasts, especially *Saccharomyces cerevisiae*, have been studied a great deal. In all the studies two important points emerge, firstly that the cell wall region is of prime importance, and secondly that the factors involved and the way they influence aggregation seem to be common to all cell types. This suggests that there may be similar mechanisms involved in the flocculation of many different cells.

2.2.1 Bacterial Aggregation

Attempts to understand the mechanism of bacterial aggregation have been mainly concerned with investigations on bioflocculation as it occurs in the activated sludge process of the biological waste water treatment industry. Lately the bacterial strains used for single cell protein production have received consideration. Flocculation can be allowed to occur naturally, or enhanced by the use of synthetic polymers. Investigations on the mechanism of flocculation under the latter conditions are carried out on the assumption that they will lead to increased understanding of natural flocculation.

It was demonstrated by Hodge and Metcalf (1958) that bacteria, which are negatively charged, can be aggregated by synthetic polymers. Cationic, anionic and non-ionic polymers were all found to be effective though in the case of non-ionic or negatively charged polymers it was necessary for multivalent ions, expecially calcium, to be present. It was suggested that sorptive interactions took place between the polymers and the surface of the bacteria followed by inter-cell bridging, i.e. the bridging mechanism.

The presence of natural polymeric materials at bacterial surfaces suggests that natural bacterial flocculation may be the result of polymer-bacteria interactions similar to those of the anionic and non-ionic synthetic polymer aggregation of bacteria.

As a result of studying the flocculation of strains of *Nocardia corallina*, Clark (1958) suggested that is is not the cell wall *per se* that is important but secretions at the cell wall-slime layer interface. He also suggested that flocculation was not caused by incomplete cell cleavage. Stanley and Rose (1967) suggested that the clumping of *Corynebacterium xerosis* appeared to be caused by a component located either at the cell surface or in the microcapsule. They claimed that the clumping of several different microorganisms has been shown to be associated with different types of cell surface components including carbohydrate, protein, and hyaluronic acid. The ability of proteolytic enzymes to decrease and even remove the clumping ability of *C. xerosis*, together with the inability of lipid solvents, enzymes, and reagents, which react specifically with carbohydrates, hyaluronic acid, mucopeptides, or sterols, to affect this ability, indicated that a cell surface protein is involved in clumping. The particular compounds involved in clump formation were suggested to be the cell-surface phosphates. The data of Stanley and Rose (1967) also indicate that α-carboxyl groups may contribute to the "stickiness" of the cell-surface component, since clumping ability was greatest around a pH of 3, which is the pK value of the α-carboxyl group. An hypothesis concerning the mechanism was put forward to the effect that, in liquid cultures or suspensions, gas bubbles provide an interface at which the bacteria collect. This was described as an "hydrophobic effect" dependent on the presence of exposed hydrophobic aminoacid residues in a cell surface protein. Renn (1956) also suggested the accumulation of bacterial mass at air-water interfaces and the subsequent formation of large floc particles. This is similar to the translocation and concentration of organic materials at the air-water interface in waters containing low concentrations of organics, which is attributed to energy loss through the work done in decreasing the interfacial tension. Bacterial aggregation in sea water has been attributed by Barber (1966) to the deposition of bacteria around air bubbles. Studies on the flocculation of mixed microbial cultures have resulted in a suggested mechanism similar to that given above for pure bacterial cultures. Pavoni *et al.* (1972)

in an exhaustive investigation of bacterial exocellular polymers and biological flocculation proposed a mechanism in which flocculation results from interactions of high molecular weight exocellular polymers, accumulated at the microbial surface during endogeneous growth. These polymers combine, either electrostatically or physically, and subsequently bridge the cells of the dispersion into a three-dimensional matrix of sufficient magnitude to settle under quiescent conditions. These workers also found that surface potential reduction was not necessarily a prerequisite for aggregation. In addition to the extracellular polymer bridge Parker *et al.* (1971) postulated that particle bonding in the activated sludge floc is also due to a filament network that provides structure for the buildup of particles which are retained by enmeshment and polymer bridging. McKinney (1956) explains the mechanism of flocculation in the activated sludge process as being primarily energy-dependent. Electrophoretic measurements made on 72 bacteria showed that although they all had surface charges (zeta potential) below the "critical" charge (Fig. 3), flocculation did not occur in all cases. Evidently there is more

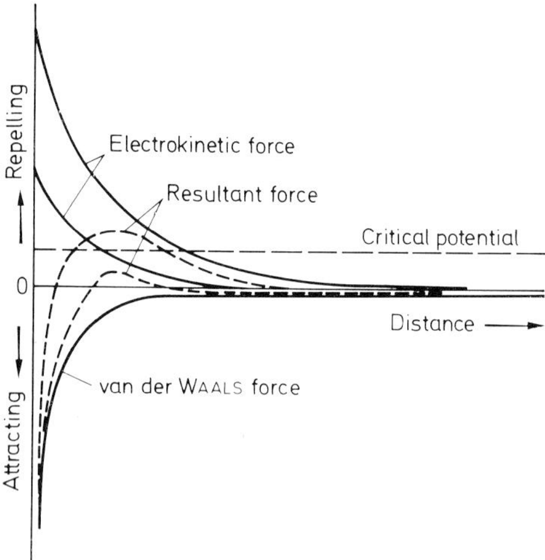

Fig. 3. Forces acting on lyophobic colloids (McKinney, 1956)

to floc formation than surface charge. It was observed further that flocculation did not result until the bacteria entered the endogenous phase of metabolism. This is explained as being due to the bacteria having insufficient energy under these conditions, to break away from the floc, which in turn permits colloidal surface forces acting on the bacterial cell to predominate and hold the bacteria to the floc. Thus the net balance of forces as two bacteria approach on a collision path is favourable to flocculation.

The other forces that might be involved in flocculation are salt linkage, direct chemical bonding, and/or physical forces. The normal thickness of the electrical double layer on the bacterial cell in waste disposal systems is less than 5×10^{-3} µm. The site of the charge at the cytoplasmic membrane would place the electrical double layer well within the cell wall rather than at the contact surface, i.e. the capsular layer (Fig. 4). In view of this

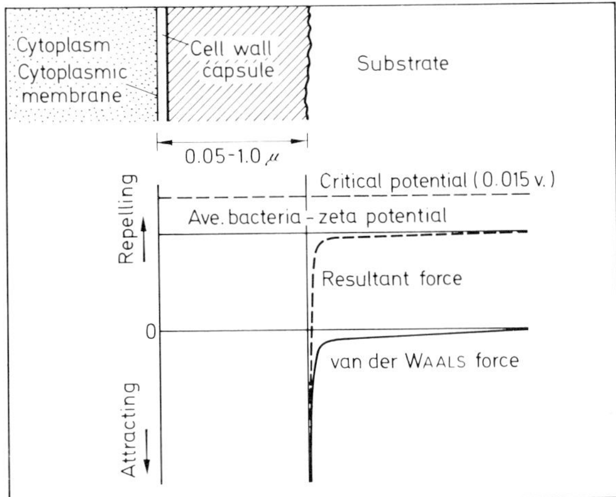

Fig. 4. Surface forces acting on bacterial cells (McKinney, 1956)

geometric factor and since linkage can occur only if the special configuration of the active groups permitted interaction of adsorbed cations salt linkage is not considered important. Chemical bonding is also considered negligible because of the low chemical reactivity of the polysaccharide capsular layers.

McKinney (1956) found support for the above views by growing *Serratia marcescens* in the dispersed state at a high organic substrate concentration, followed by separation of the cells from the substrate, washing the excess organic matter from the surface of the cell, and then resuspending the cells in normal salt solutions. The last step brought about flocculation, indicating that bacteria were flocculated as a result of a low energy level and subsequently held as floc by physical forces.

The presence of spherical bodies in the flocs of a protozoa-bacteria culture was found to result from the death of protozoa and their subsequent lysis; followed by the liberation of spherical bodies. With species of bacteria which formed flocs at a relatively slow rate, the protozoon *Tetrahymena gelta* caused floc formation to proceed more rapidly. A similar observation was made by Parker *et al.* (1971).

2.2.2 Brewers' Yeast

A great deal of work has been carried out in studying the flocculation of the brewers' yeast *Saccharomyces cerevisiae.* This has been developed to the extent of determining the chemical composition of the yeast cell wall and the components responsible for flocculation. Mill (1964) proposed a mechanism based on the presence of calcium ions in the medium. The calcium atom forms a bridge between receptor sites, possibly carboxyl groups, on different yeast cells. The bonds are at first essentially ionic, but once established the cells transitorily approach closely to one another and hydrogen bonding is set up between complementary carbohydrate structures in the walls of the two cells. The resulting complex thus assumes a chelate character, with the calcium-complexing groups held in a definite spatial relationship to each other. In determining the wall components Mill (1966) found that there was little difference in the polysaccharide composition of the walls of flocculent and non-flocculent strains of *Saccharomyces cerevisiae*, but the degree of phosphorylation of the mannan was greater in flocculent than non-flocculent strains. The observation (Mill, 1964) that certain nitrogeneous compounds when added to the medium delayed the onset of flocculation was related (Mill, 1966) to the fact that a higher nitrogen content was found in the non-flocculent walls. Lyons and Hough (1970, 1971) confirmed Mill's results and showed that the flocculent yeast cell wall contains a higher level of phosphate groupings in the outer layer of the mannan-protein. The level of phosphate in the outer layer is sufficient in the case of flocculent strains to enable binding of bivalent cations between adjacent cells. The walls of flocculent yeasts bind on average twice as much calcium as walls of non-flocculent yeasts and removal of the phosphorylated manno-protein decreased the capacity of the cell wall to bind calcium and rendered the cell non-flocculent. Lyons and Hough (1970) also showed that there was sufficient phosphate present at the surface of flocculent yeast to enable one calcium molecule to be bound by two phosphate units. If this calcium were bound by phosphate from adjacent cells then flocculation might occur. In non-flocculent strains there is insufficient phosphate at the surface.

2.2.3 Fungal Pellets

The type of growth of filamentous fungi in submerged culture varies from the "pellet" form, consisting of compact discrete clumps of hyphae, to the "filamentous" form, in which the hyphae form a homogeneous suspension dispersed through the medium. The importance of *Penicillium chrysogenum, Aspergillus niger* and *Agaricus campestris* in the production of penicillin, citric acid and mushrooms, has led to the study of their pelleting behaviour and its effect on the various processes. The mechanism postulated differs markedly from that encountered in the flocculation of bacteria and yeasts. Foster (1949) found that the development of the aggregate colony type of growth occurred when the nutrient conditions were unfavourable for rapid and abundant growth, and where agitation conditions were very mild. He regarded a small number of inoculum particles to be the most important factor in achieving such growth. Foster suggested that spores germinate at different rates and those germinating first, forming germ tubes and secondary hyphal branches, probably trap other ungerminated or germinated spores and form the nucleus of the large pellet type of colony. Whitaker and Long (1973) in a recent review of fungal pelleting, suggest that the process of pellet formation can be divided into two stages: 1. the aggregation of spores at an early stage of germination when the spores are swelling; 2. the aggregation of small clumps of germinated spores. Trinci (1970) has observed the aggregation of spores of *P. chrysogenum* before the onset of germ tube formation. The formation of a trap-net by the aggregation and interlocking of germ tubes and hyphal branches proposed by Foster has been confirmed by Pirt and Callow (1959) (Fig. 5). They observed that the initiation of pellet formation seemed to be enmeshment of the hyphae within the individual fragments of mycelium. Block *et al.* (1953) found, with continued sub-culturing, a mutant strain of *Agaricus blazei* which grew in a filamentous form whilst the parent strain formed pellets. They attributed the filamentous and rapid growth of the mutant to the germination and

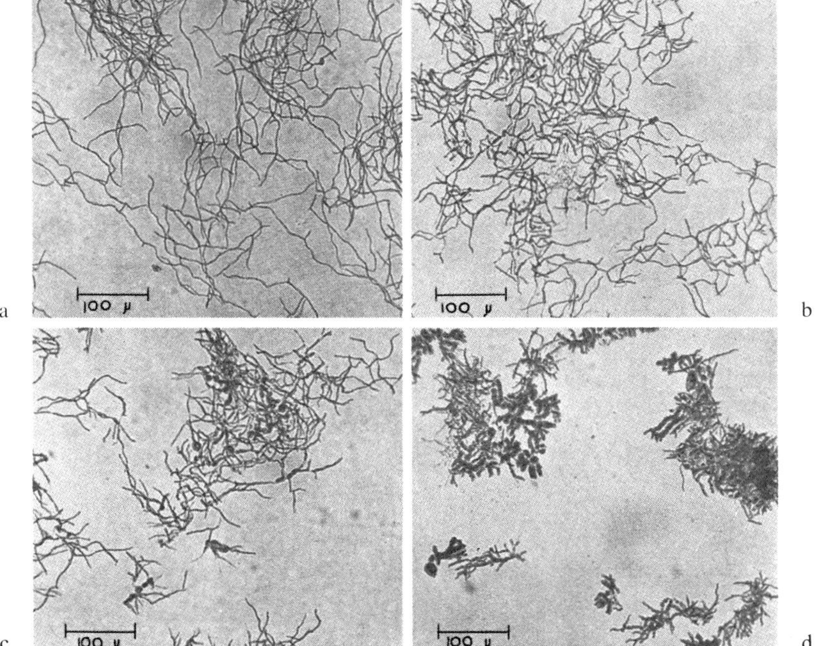

Fig. 5 a–d. Mycelial forms of *P. chrysogenum* strain Wis 54–1255 in cultures grown at different pH values. (a) pH 6.0; (b) pH 6.5; (c) pH 6.9; (d) pH 7.4 (Pirt and Callow, 1959)

growth of "secondary spores". After about 48 hrs the mycelial strands began to fragment and contributed to increased growth with each individual fragment serving as a nucleus for new growth.

It is clear from the above discussion that the mode of growth of the fungi ultimately determines the morphology of the culture.

2.2.4 Plant-Cell Culture

In a review on plant-cell culture, Puhan and Martin (1971) claim that the proportion of free cells to aggregates, and the average size of the cells and cell aggregates, vary with time, the composition of the medium, and the age of the culture. However, the ideal conditions for cell division followed promptly by cell separation, i.e. a culture composed of completely free cells, has not so far been achieved. Furthermore, no procedure has yet been devised for separating plant cells comparable with the trypsinization process used with animal-cell cultures. The use of chemical and enzymic methods causes injury or death to the plant cells. Mandels (1972) observed that plant cells grow as large masses adhering to baffles, above the liquid line, in the foam, or clogging air inlets and outlets, harvest and feed lines. Unlike bacteria, plant cells do not separate after cell division in the intact plant and this tendency persists in cell cultures. On the origin and development of embryoids in suspension cultures of carrot, McWilliam *et al.* (1974) conclude that embryoids are initiated at or near the surface of characteristic cellular aggregates (embryonic clumps) and are released from these aggregates as free-floating structures capable of further development into plantlets. They observed that the superficial cells of the clumps were clearly distinguishable from the central cells and contained large starch grains, a large central nucleus, abundant cytoplasm and a number of vacuoles. The embryoids were seen to arise from individual superficial cells.

2.2.5 Cell Tissue

In discussing the phenomena of cell association Moscona (1961a) gives sponge as an example. When cells of different sponge species are dispersed and then mixed together, they separate and reaggregate by species, forming separate clusters. The cells are able to identify one another and so associate preferentially with their kin. In his work on the aggregation of dissociated embryonic cells Moscona found that the cementing substances of mammalian tissues yielded to digestion by trypsin, and certain other enzymes that break down proteins without serious injury to the cells and that the dispersed cells of mammalian or bird embryos will readily aggregate into clusters. Cells from different kinds of tissue were observed to sort themselves out by cell type in forming these clusters. The intercellular cements have remarkable flexible and dynamic properties. Although they bind the cells they permit them to move about and regroup without actual loss of contiguity. This dynamic linking is a cardinal feature of cell contact at all levels of multicellular organization.

As a result of studying the effect of temperature on adhesion to glass and histogenetic cohesion of dissociated cells, Moscona (1961b) suggested that cell adhesion depends on metabolic sequences and closely reflects the metabolic economy of the cells and is not merely a result of coupling between stable surfaces, but also a function of the synthetic activities of cells. The mechanism proposed for adhesion and histogenetic bonding of cells involves the production and organization between cells of mucoprotein-containing extracellular materials, the formation or effective function of which requires divalent cations and certain large molecules in the cellular environment (Moscona and Moscona, 1966).

Gierer (1974) used isolated cells and regenerating tissues of hydras (multicellular organisms about half a centimetre long) as a model for cell differentiation and pattern formation, and found that dissociated hydra cells first form aggregates, or clumps, before regenerating into complete animals. Gierer also found that isolated cells remained alive and tended to aggregate if they were placed in solutions of high salt concentration in which whole animals slowly disintegrate. The evidence from observations of the aggregation and regeneration in hydras suggested that the invisible spatial pre-patterns that precede the patterns of development are embodied in chemical substances. If a specific organic substance, that occurs naturally in the animal, affects morphogenetic processes at low concentrations, it is probable that the same substance is also involved in natural morphogenesis. Gierer (1974) isolated, from hydra tissue, activating and inhibiting substances that influence budding, and subsequent regeneration of the animal. He proposed a molecular model for morphogenesis; the postulate being that there exist activator molecules, inhibitor molecules and a receptor protein that is activated by association with two activator molecules. Activators and inhibitors are assumed to be contained separately in small storage sacs but can be released by the special protein in its activated form. Once released, the activator and inhibitor are in the active state and are free to diffuse through the surrounding tissue. The inhibitor can prevent the release of activator by occupying release sites on the activator storage sac but cannot block the release of inhibitor. The inhibitor molecules are assumed to diffuse more freely and thus have a wider range between release and degradation than the activator molecules. The resulting distribution of activator directs the generation of patterns with the desired properties. Gierer also attributes some regulatory effects to genes being turned on or off; others may be due to the enzymic synthesis of morphogenetic substances.

3. Factors Affecting Floc Formation

The activity of the single cell and its interplay with its environment constitute the primary element of a microbiological system. The actions of the individual cell are genetically controlled and consequently the chemical structure of each cell dictates the response to a particular environment. The perpetual interaction and interdependence of the total environment and the living cell leads to the maintenance of a balanced existence within each system, and preserves coexistence between the surviving environment and organisms. The factors that influence flocculation obviously fit in with this overall purpose. Microbiological and environmental factors, as well as nutrition, directly affect the metabolism of the living organism. The cell-wall region, regarded as the crucial area responsible for aggregation, is influenced by microbiological and environmental factors directly and indirectly through metabolism. The environment plays a role in deciding the condition of this region as well as the general character of the aggregate, while the aggregate by its formation and presence in the system, modifies both the metabolism of the cell and the environmental conditions (Fig. 6). The particular biochemical process dictates the importance or otherwise of certain factors on flocculation and in turn on the process.

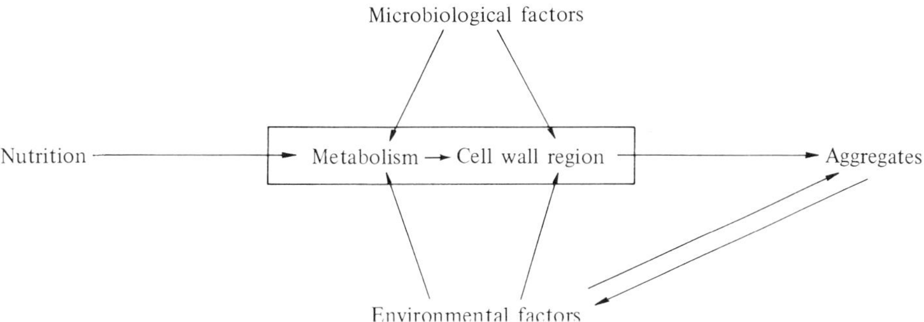

Fig. 6. Factors Affecting Aggregation

Table 3 gives a list of the factors that affect flocculation, some of which are of common importance to all living cells while others are of a particular relevance only to a certain genus or type.

Table 3. Factors affecting aggregation

Microbiological	Environmental		
	Physical	Chemical	Biological
Cell wall	Hydrodynamic properties	Presence of	Inoculum size
Genetics	aeration	chelating agents	Presence of other
Growth rate	dilution rate	C/N ratio	organisms or strains
Nutrition	fermenter size	enzymes	
Physiological age	time and intensity	ferrocyanide	
	of agitation	nitrogeneous	
	viscosity	substances	
		oils	
	Interfacial phenomena	sugars	
	gas bubbles	trace metals	
	surface active agents		
	suspended solids		
	Ionic properties		
	organic solvents		
	pH		
	presence of calcium ions		
	Temperature		

3.1 Microbiological Factors

3.1.1 Genetics

There is evidence that flocculation is genetically controlled and thus may be an inherent property in the racial character of the microorganism.

The presence of zoogloea-producing bacteria has been found not to be necessary for the formation of flocs in the activated sludge process. McKinney (1956) lists a number of floc-producing organisms which have been isolated from activated sludges (Table 4). Isolation studies carried out on sludges obtained from various domestic, synthetic and industrial sources yielded 72 bacteria of which there were 10 species capable of forming flocs similar to activated sludge, within a 48-hrs aeration period. Stanley and Rose (1967) list the species and strains of coryneform bacteria which have the ability to flocculate.

In bottom fermentation for beer production a yeast is chosen which has good flocculation properties. Thorne (1951) has shown that flocculence is a genetically controlled

Table 4. Floc-producing bacteria (McKinney, 1956)

Organism	Habitat
Alcaligenes faecalis	intestinal canal
Alcaligenes metalcaligenes	intestinal canal
Bacillus cereus	soil
Bacillus lentus	soil
Escherichia coli	intestinal canal
Escherichia freundii	soil
Escherichia intermedium	soil
Flavobacterium sp.	water
Nocardia actinomorpha	soil
Paracolobactrum aerogenoides	soil
Pseudomonas fragi	soil
Pseudomonas ovalis	soil
Pseudomonas perlurida	soil
Pseudomonas segnis	trickling filter
Pseudomonas solaniolens	soil
Zoogloea ramigera	activated sludge

character of *Saccharomyces cerevisiae* and that mutation leads to flocculent strains. The importance of flocculation to the brewer has led to most studies being confined to brewers' yeast. Analogous phenomena have been shown to exist in the fission yeast *Schizosaccharomyces pombe* (Calleja, 1970).

In a review on fungal pelleting Whitaker and Long (1973) presented information summarising the presence or otherwise of pelleting in a large number of species from a range of genera (Table 5). Pirt and Callow (1959) found that the various strains of *Penicillium chrysogenum* differed in their ability to form pellets.

In his work on the aggregation of embryonic cells Moscona (1961a) found that aggregation patterns vary with different types and mixtures of cell. Under otherwise identical conditions, different kinds of cell "crystallized" into distinct and characteristic aggregates, varying in shape and size. Remarkably, those patterns that showed themselves to be characteristic of particular kinds of tissue proved to be similar for cells from different species. Whether from mouse or chick embryo, cells of the same tissue aggregate into very similar patterns. Their collective reactions seem to be guided by factors identifiable by both species.

3.1.2 Nutrition

Mill (1964a) carried out experiments in which freshly harvested yeast was added to solutions of glucose, yeast extract and ammonium salt in various combinations (Table 6). No flocculation occured in the absence of glucose, but developed rapidly in the presence of glucose alone, or glucose and yeast extract, but was much delayed when ammonia was also present.

Pirt and Callow (1959), in attempting to control the morphology of *Penicillium chrysogenum*, observed that in addition to pH nutrition affected hyphal length and morphol-

Table 5. Pellet -forming fungi (Whitaker and Long, 1973)

Phycomycetes	Basidiomycetes	Fungi imperfecti
Mucorales	*Polyporales*	*Moniliales*
Absidia glauca	*Cantharellus cibarius*	*Aspergillus candidus*
A. spinosa	*Fomes pinicola*	*A. clavatus*
Chlamydomucor javanicus	*Lycoperdon umbrinum*	*A. flavicepes*
Cunninghamella bainieri	*Merulius tremellosus*	*A. flavus*
Phycomyces blakesleeanus	*Polyporus sulphureus*	*A. giganteus*
Rhizopus acidus	*Poria subacida*	*A. nidulans*
R. chinensis	*P. undora*	*A. niger*
R. oryzae	*Trametes heteromorpha*	*A. ochraceus*
R. Peka 11	*T. seriales*	*A. stellatus*
Entomophthorales	*Agaricales*	*A. terreus*
Basidiolus ranarum	*Agaricus bisporus*	*A. versicolor*
Conidiolus sp.	*A. blazei*	*A. wentii*
	A. campestris	*Cephalosporium ulmi*
Ascomycetes	*A. rodmanni*	*Dactylium dendroides*
	Collybia umbulata	*Fusarium oxysporum*
Pezizales	*C. velutipes*	*F. solani*
Morchella conica	*Hebeloma sinapizans*	*Gliocladium roseum*
M. crassipes	*Hypholoma fasiculare*	*Macrosporium sarcinaeforme*
M. esculenta	*Lentinus elodes*	*Myrothecium verrucaria*
M. hortensis	*Lenzites trabea*	*Penicillium brevicaule*
M. rimosipes	*Lepiota naucina*	*P. camenbertii*
M. rotunda	*L. procera*	*P. chrysogenum*
M. semilibera	*Pleurotus corticatus*	*P. clavariforme*
M. vulgaris	*P. ostreatus*	*P. digitatum*
Morchella sp.	*Psalliota campestris*	*P. italicum*
Hypocreales	*Psilocybe sp.*	*P. notatum*
Gibberella fujikuroi	*Schizophyllum commune*	*P. roquefortii*
G. saubinetti		*P. urticae*
Sordaria fimicola		*Trichoderma viride*
Sphaeriales		*Sphaeropsidales*
Neurospora crassa		*Alternaria solani*
Chaetomium caprinum		*A. tenuissima*
C. olivaceum		
Pseudosphaeriales		
Guignardia bidwelli		

ogy. They attribute pelleting to the formation of swollen cells and hyphal branching, and found that substitution of the nitrogen source (ammonium sulphate), by corn-steep liquor, reduced the frequency of hyphal branching and prevented the formation of swollen cells. This strongly suggests that corn-steep liquor contains some substance which stimulates the production of the long thin hyphal type. A similar effect was found by Clark (1962a) in that pellets of 3.2 mm diameter produced by *A. niger* in cane molasses were approximately twice the size of those produced in beet molasses.

Table 6. Flocculation of yeast cells incubated in various mixtures of media components (Mill, 1964a)

Components present in (+), or absent from (–), the incubation mixture				
Glucose (1.25%)	Yeast extract (0.25%)	Ammonium succinate (0.06 M)	Flocculence as indicated by Log (Sedimentation rate)	
			After 4 hrs	After 24 hrs
+	–	–	2.37	2.82
+	+	–	2.47	2.92
+	+	+	1.42	2.05
+	–	+	1.42	1.70
–	+	+	1.32	1.20
–	+	–	1.32	1.20
–	–	+	1.20	1.20
–	–	–	1.27	1.20

All the mixtures of media components contained 10% v/v of salts solution. The cells had a log sedimentation rate of 1.20 before incubation.

3.1.3 Physiological Age (Growth Phase)

Stanley and Rose (1967) have shown that the clumping ability of bacteria increases with the age of the culture from the mid-exponential to the late-exponential phases of growth, but declines as the culture enters the stationary phase (Fig. 7). Similarly,

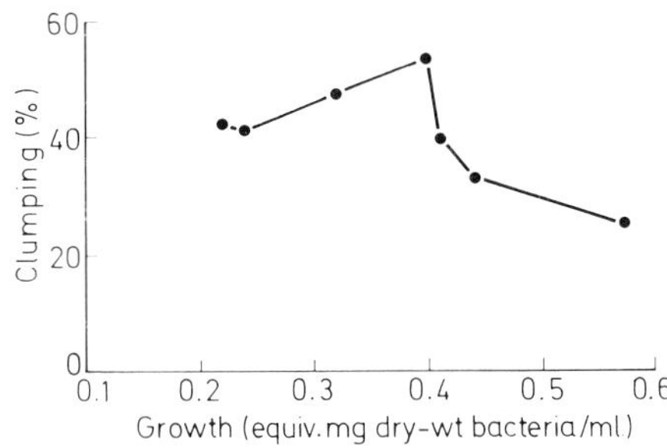

Fig. 7. Effect of culture age on clumping of *Corynebacterium xerosis* NCIB 9956 at 15° and pH 7.0. Observations were made during the late-exponential phase (equiv. 0.2–0.4 mg dry-wt/ml.) and early stationary phase (equiv. 0.4–0.55 mg. dry-wt/ml.) of cultures. Bacteria were harvested from cultures grown at 30° and the extent of clumping measured (Stanley and Rose, 1967)

Pavoni *et al.* (1972) obtained a connection between bacterial growth phase and bacterial flocculation (Fig. 8), which indicated that aggregation was governed by the physiological state of the microorganisms. Flocculation was not observed to occur until the microorganisms entered the endogeneous phase. The data curves obtained when using organics as a sole carbon source were of the same shape as those for glucose.
During the fermentation of malt wort by *Saccharomyces cerevisiae* Baker and Kirsop (1972) found that the ability to aggregate diminished during exponential growth and intensified once more as this stage terminated. Mill (1964a) found the same relationship between growth phase and flocculation (Fig. 9); he also claimed that the less flocculent the yeast cells the more rapidly they grew.

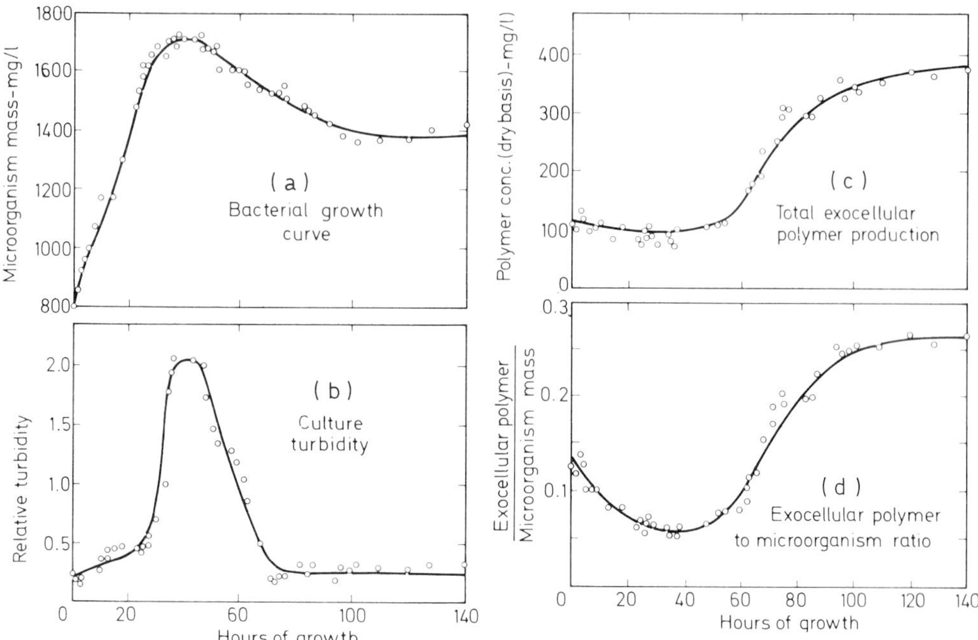

Fig. 8 a–d. Relationship between growth phase of mixed bacterial culture (a) and bacterial flocculation (b). Also shown are the accumulation of exocellular polymer (c) and the ratio of accumulated exocellular polymer to microorganisms (d) (Pavoni *et al.*, 1972)

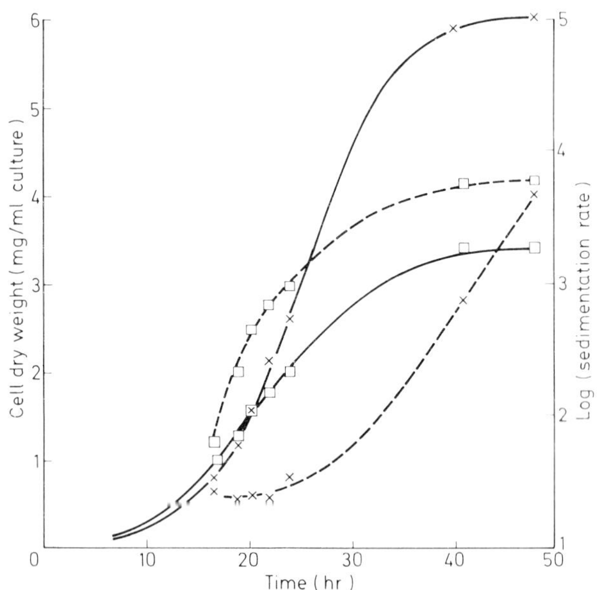

Fig. 9. Growth (solid lines) and flocculation (broken lines) of *Saccharomyces cerevisiae* in GPF medium, □, GPF medium + 0.012 M-ammonium succinate, ×. Log SR = log equivalent μg. dry wt. organism/ml. min (Mill, 1964a)

The above observations are generally supported by the work of Parker *et al.* (1971) who found that flocs were more susceptible to erosion at higher than lower substrate removal rates.

Cocker and Greenshields (1975) have studied the morphological development of *A. niger* and have also found that culture age has an influence on morphology. In the early phases of growth the morphology depended largely on spore concentration and varied from filamentous growth, to loosely and densely packed pellets. When the stationary phase was reached, increased autolytic activity coincided with increased heterogeneity of appearance of individual particles.

3.1.4 Cell-Wall

The importance of the cell-wall region in bacterial and yeast flocculation has been emphasised by many workers (Eddy, 1958; Hough *et al.*, 1971), who isolated cell walls and demonstrated that their flocculation characteristics were related to those of the corresponding whole cells. It follows that the composition, structure and charge of the wall and the presence of extracellular polymeric materials are factors to be considered. In contrast the mode of flocculation of fungal organisms involves mycelial growth and its enmeshment followed by entrapment of spores and mycelial fragments.

A great deal of work has been done to establish the composition of the yeast cell wall (Fig. 10), particularly of brewers' yeast, and the relationship between composition and

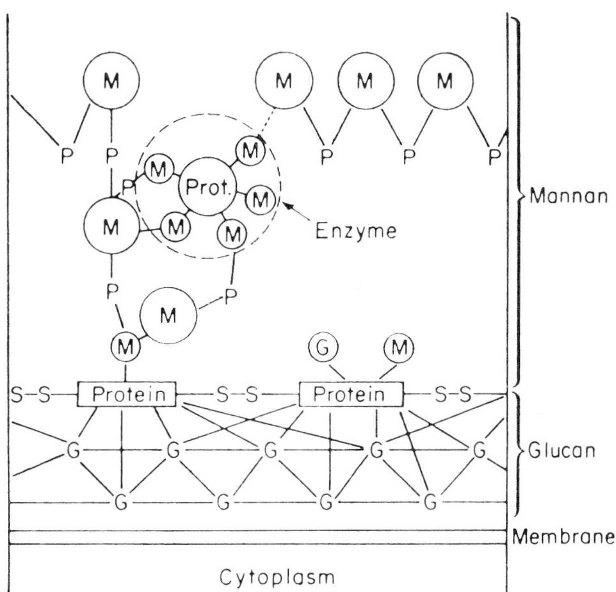

Fig. 10. Schematic structure of yeast cell wall. M–Mannan, P–Phosphate, G–Glucan, S–Sulphur (Hough *et al.*, 1971)

flocculation (Section 2.2.2). Masschelein and Devreux (1957) showed that flocculent yeast cell walls contained a calcium-binding component and had a high phosphorus content. They concluded that the outer cell wall is a mannan-phospholipid-protein complex. Eddy and Rudin (1958) proposed that yeast cell walls consisted of a solid glucan matrix to which glucan and mannan are bound through protein. The mannan also seems to carry phosphate groups. The protein's carboxyl groups may be responsible for the change in charge from positive to negative over the range of pH 3–7. They regarded flocculation not to be simply related to cell charge, but, since papain solubilises the protein-mannan complex with loss of flocculence, the cell-wall constituent carrying the charge is involved in flocculation.

Pavoni *et al.* (1972) have recorded electrophoretic mobility throughout an entire bacterial growth cycle (Fig. 11). The data depict a constant bacterial surface charge throughout

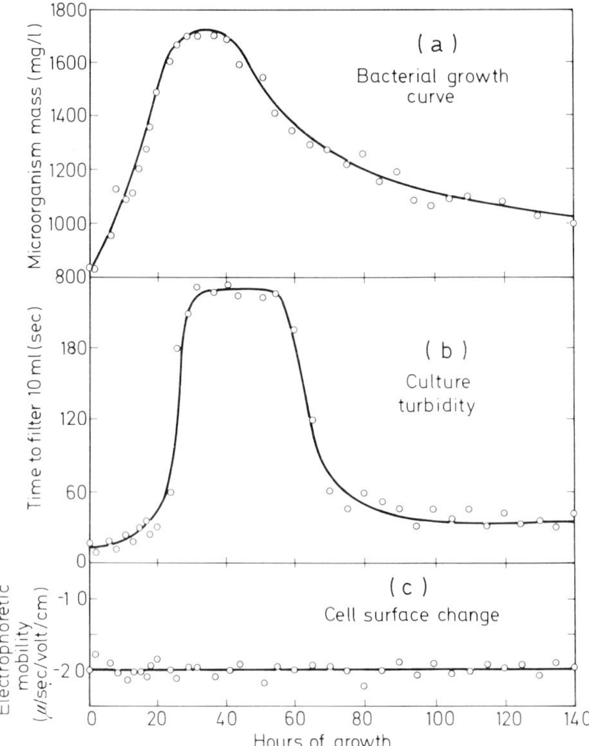

Fig. 11 a–c. Relationship between growth phase of mixed bacterial culture (a), biological flocculation (b), and microbial surface charge (c) (Pavoni *et al.*, 1972)

all growth phases regardless of the degree of flocculation of the culture. They did not consider surface charge reduction to be a necessary precursor to biological flocculation. Similarly McKinney (1956) has found that many bacteria which had surface charges below the "critical" charge for bacterial flocculation, did not flocculate (Section 2.2.1). Electrophoretic mobility studies on strains of top and bottom fermenting yeasts, at pH 3–7, have shown that the mobilities were independent of the flocculation characteristics of the yeasts (Eddy and Rudin, 1958). This indicates that flocculence and surface charge of the yeast cell are not interrelated.

Lindquist (1953) proposed that the cell wall contains both positive and negative ionogenic groups. Under favourable conditions, these polar groups could attract oppositely charged groups on other cell surfaces and flocculation would result. It is not a low level of net charge which is essential, but accessibility of the positive ionogenic groups, which may be masked, perhaps by protein.

The presence of extracellular polymers has also been implicated in flocculation. Macromolecules are secreted or exposed at the outmost surface of microorganisms. Pavoni *et al.* (1972) found that there was a direct correlation between exocellular polymer accumulation and microbial aggregation, the major compositional make-up of the bacterial polymers being polysaccharide, protein, RNA, and DNA. Removal of exocellular polymeric material from well-flocculated bacterial suspensions followed by subsequent resuspension of the harvested cells, resulted in stable dispersions. Subsequent addition of extracted exocellular polymer resulted in reflocculated suspensions.

The presence of large polymer-like molecules at the wall of yeast cells is of a different nature to their presence in bacteria. They are more a part of the wall than being extracellular. Harris and Mitchel (1973), in a comprehensive review on the role of polymers in aggregation, state that while glucan, mannan, chitin, protein, and lipid are all found in the cell wall of all brewing yeasts, the phosphomannan-protein present on the outer surface of the cell wall seems to be most intimately involved in flocculation since removal of this layer by treatment with alkali, trypsin or papain decreased flocculence. As mentioned previously the mechanism of flocculation involves cross-bridging, *via* calcium ions, between the phosphomannan-protein in adjacent cell walls (Section 2.2.2).

3.2 Environment

Environmental factors affecting flocculation (Table 3) have been divided into physical, chemical, and biological groups rather than on the basis of their action and its influence on flocculation, since their effect may be through one or more mechanisms. Some environmental factors, such as agitation, exert an action on the aggregates at all times, from formation, through the presence of stable flocs, to the stage of breakup. Other factors however may only influence either the micro-structure or the macro-structure of flocs but not necessarily both. The action of an individual factor in the process of flocculation may be direct or indirect, for example, a certain factor may by its presence interact directly with the cell wall, causing its modification, or directly participate in the cell-cell interaction. Other factors may, through changing cell metabolism, indirectly affect the cell-wall region. Another important point is that of the reversibility of the influence of a certain factor on flocculation. Depending on the nature of the action of the factor some effects may be reversed, while others cause permanent changes.

3.2.1 Temperature

Stanley and Rose (1967) found that the clumping ability of bacteria harvested from exponential-phase cultures and suspended in buffer increased as the temperature of the suspension was decreased below 30° C. The extent of clumping at any one temperature depended on the pH value of the buffer. At pH 7.0, clump formation was not measurable at 30° C, but at pH 3.0 it was still appreciable at 40° C (Fig. 12).

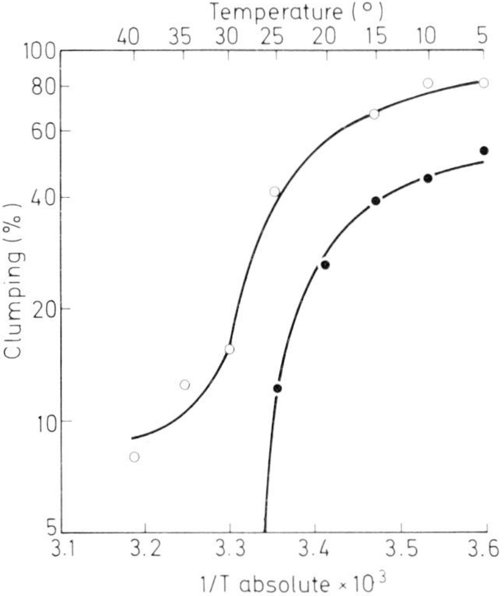

Fig. 12. Arrhenius plots for the clumping of *Corynebacterium xerosis* NCIB 9956 at pH 3.0 (○; HCl + KCl; Bower and Bates, 1955) and pH 7.0 (●; Na_2HPO_4 + NaH_2PO_4; Gomori, 1955). Bacteria were harvested from cultures containing equiv. 0.2–0.3 mg. dry-wt bacteria/ml. The extent of clumping at the different temperatures was then measured (Stanley and Rose, 1967)

Mill (1964a) observed that, when a suspension of flocculent yeast in calcium chloride and sodium acetate buffer was gently heated from 23–62° C, there was at first a slight decrease in flocculation as the temperature was increased while a rapid change occured between 50° and 60° C. At 60° C the yeast was virtually completely dispersed, but as the temperature was lowered once more, the flocs reappeared. The effect of temperature on the stability of flocs of *Schizosaccharomyces pombe* in de-ionized water has been shown (Calleja, 1970) to be similar to that reported for brewers' yeast (Mill, 1964a). Flocs were stable in the range 20–50° C; higher temperatures caused deflocculation, which was complete at 80° C and above.

Studying the morphological development of *A. niger* during tower fermentation Cocker and Greenshields (1975) observed that the colonies varied in appearance according to the branching frequency and packing density of their hyphae. They classified the colonies into seven representative reference types (Section 4.1.1). Fermentations at 25° C produced colonies of 1–20 mm diameter flocs, with a tendency toward a growth type in which the hyphae develop radially and evenly to form spherical colonies with low hyphal packing densities. At 35° C, growth led towards colonies of a loose feathery type.

A decrease in temperature seems to favour flocculation of microorganisms while the opposite was observed by Moscona (1961b) in the aggregation of embryonic cells (Fig. 13). Between 30° and 25° there was a significant and consistent decrease in the size of aggregates of all cell types tested. At 15° C the cells did not cohere at all and aggregates were not formed, this indicates that simple reactions between divalent cations

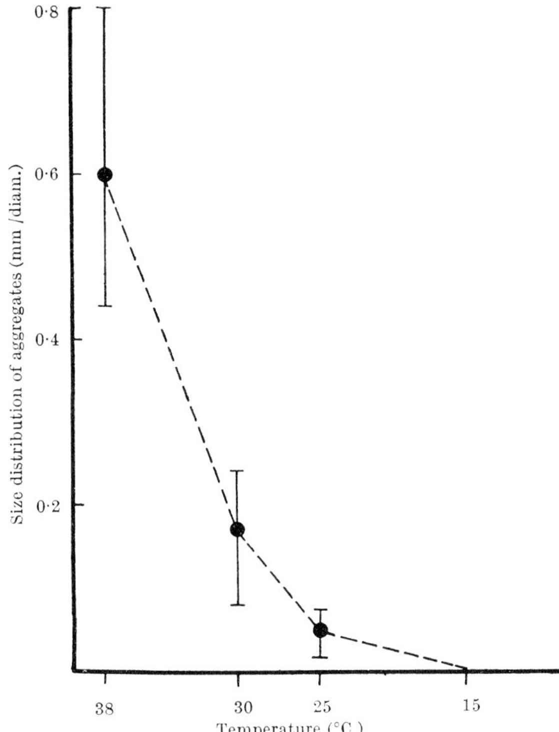

Fig. 13. Effect of temperature on size distribution of aggregates of dissociated neural-retina cells (Moscona, 1961b)

(calcium, magnesium) or other first-order physico-chemical effects cannot be exclusively responsible for cell adhesion, as such reactions are unlikely to be persistently inhibited at 15° C.

3.2.2 Ionic Properties

The mechanisms of aggregation which have been proposed involve ionic phenomena, the ionic properties of the solution being found to play an important role. These properties include the level of hydrogen ions (pH), and the presence of polyvalent ions and organic solvents.

It has been emphasised by many workers that pH changes seem to have a profound effect on flocculation. For example, the clumping ability of *C. xerosis* was shown to be greatest when the bacteria were suspended in buffer at pH 3.0 (Stanley and Rose, 1967), while between pH 3 and 5, there was a decline in the percentage of bacteria that clumped, but clumping ability was little affected by pH values between 5 and 10 (Fig. 14).

Pavoni *et al.* (1972) found that the pH of a suspension of colloids and bacteria was the governing factor in achieving flocculation when using extracted exocellular polymeric material (Fig. 15). Bacterial cells, whose exocellular polymer was removed, were suspended in a salt solution and their flocculation was observed over a pH range of 1.0–12.0. The experiment was repeated with the previously extracted bacterial exocellular polymer

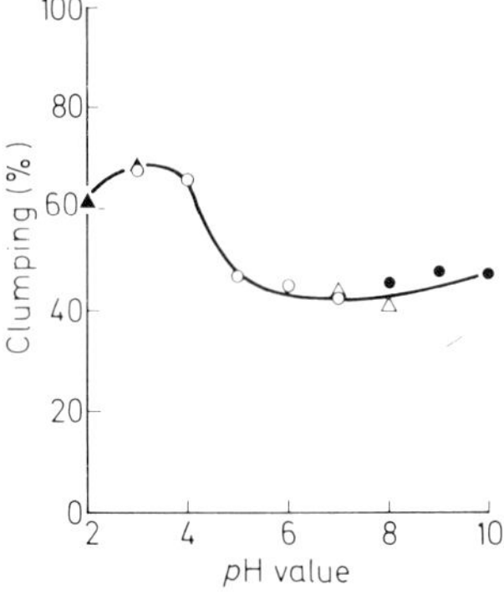

Fig. 14. Effect of pH on clumping of *Corynebacterium xerosis* NCIB 9956. The buffering systems used were HCl + KCl (▲; Bower and Bates, 1955), citric acid + Na_2HPO_4 (○; McIlvaine, 1921), Na_2HPO_4 + NaH_2PO_4(△; Gomori, 1967), and borate + KCl + NaOH (●; Bower and Bates, 1955). Bacteria were harvested from cultures containing equiv. 0.2–0.3 mg. dry-wt bacteria/ml. The extent of clumping was then measured (Stanley and Rose, 1967)

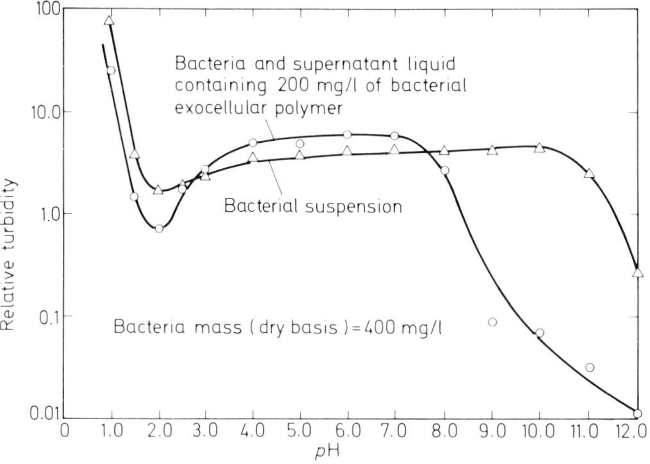

Fig. 15. Effect of pH on flocculation of mixed bacterial culture with supernatent liquid containing 200 mg bacterial exocellular polymer (Pavoni *et al.*, 1972)

re-introduced. The trend is similar to that observed by Stanley and Rose (1967); cell lysis, accompanied by natural polymer release, was postulated to be responsible for the increased flocculation in the bacterial suspension at high pH values, i.e. pH: 11.0.

In the case of brewers' yeast, Mill (1964b) observed that flocculation was low at pH 2 but rose with increasing pH and reached a maximum between pH 4.5 and 5.5.

Steel *et al.* (1954) studied the effect of pH, ferrocyanide and spore levels in relation to citric acid production. They state that of the three factors, pH appeared to have the most pronounced effect on the morphology of the inoculum. Increase in pH from 6.0 to 7.0 decreased the rate of pellet development but this could be counteracted by increasing the spore level. The extremes of pH, 5.0 and 8.0, were unsatisfactory for production of the desired seed, while at pH 4.5 or lower the spores did not germinate.

As a result of studying the influence of pH on the morphology of *P. Chrysogenum*, Pirt and Callow (1959), observed that the average length of branches decreased progressively from the order of 200 μm at pH 6.0 to 20 μm at pH 7.4. The thickness of hyphae was 2–3 μm at pH 6.0, but at 7.5, owing to the formation of swollen cells, the thickness covered the wider range, 2–18 μm. At pH 7.4 swollen, yeast-like cells were produced marking the beginning of pellet formation.

It has been reported that the presence of cations, particularly calcium, is necessary for flocculation to occur. Stanley and Rose (1967) found that *C. xerosis* did not clump in distilled water at 15° C when the suspension was rapidly stirred, but when salts were present clumps were formed. In the presence of NaCl, clump formation was greatest in solutions with an ionic strength of about 0.1. Divalent magnesium and trivalent ferric ions were also effective.

In the case of *S. cerevisiae* Mill (1964b) reports that, when potentially flocculent yeast were suspended in de-ionized water and various amounts of calcium chloride added, maximum flocculation occurred at a calcium ion concentration of about 200 μM. Two further experiments were carried out one in the presence of M-NaCl and the other with M-KCl. The NaCl antagonized the action of the $CaCl_2$ so that higher concentrations of the latter were needed to achieve any given degree of flocculation; KCl had little effect. A number of ions, for example ferric, tin and silver, have the ability to cause flocculation of a number of cells which are unaffected by the presence of calcium. These aggregates differed in appearance from those obtained with calcium ions. Rainbow (1966) in his review on flocculation reports that many workers agree on the importance of the presence of bi- and polyvalent cations, especially calcium. Calleja (1970) has shown that the same phenomenon is observed with *Schizoasaccharomyces pombe.* It is interesting to find that the presence of calcium ions was also important in the binding of embryonic cells (Moscona, 1961a).

The presence of organic solvents has also been found to affect flocculation. Mill (1964b) observed that when organic solvents were added to suspensions of non-flocculent yeast in the presence of traces of calcium chloride, flocculation occurred. On removing the solvent the yeast dispersed once more. Methanol, ethanol, propanol, acetone and dioxane were each effective when used in concentrations such that the mixture had a dielectric constant approaching 68. The addition of organic solvents has the effect of reducing the dielectric constant of the aqueous solution, and this in turn diminishes the degree of ionization of the salt bonds, increasing their strength and the contribution of the hydrogen bonds.

3.3 Hydrodynamic Properties

These properties include the time and intensity of agitation, dilution rate, aeration, viscosity of medium, and fermenter size and shape in so far as they affect the flow patterns.

In a comprehensive investigation of the physical conditioning of activated sludge floc, Parker *et al.* (1971) claimed that floc breakup, as a result of surface erosion, seems to be caused by surface shearing forces generated by fluid motion relative to the floc, and occurs when the shear strength of the bonds joining individual particles to the floc surface is exceeded. Floc breakup because of filament fracture seems to occur when excessive tensile stresses are imposed on the floc and this results in smaller flocs by fragmentation. Since the shearing stress exerted on the surface of a floc increases with floc diameter, it follows that there is a maximum stable floc size at which the imposed shear stress equals the forces binding the floc. Parker *et al.* (1971) present experimental data on particle aggregation as a function of root mean square velocity gradient, mean hydraulic residence time and inlet concentration of particles. They found (Fig. 16) that there exists a minimum detention time before any performance level is attained, and

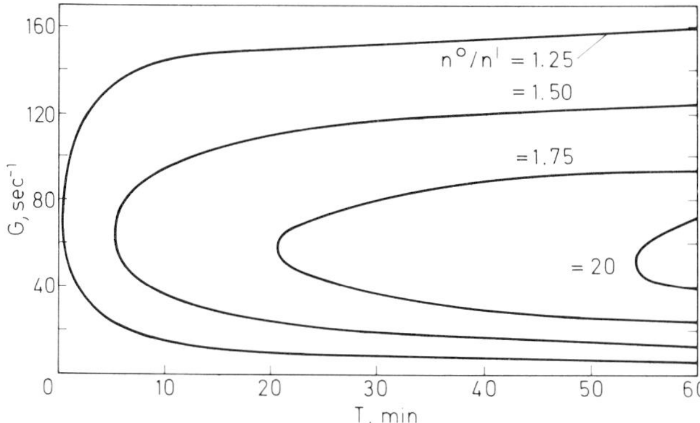

Fig. 16. Contour diagram of experimental data on the aggregation of activated sludge flocs in a stirred tank (n^1, particle concentration in effluent; G_0, rms velocity gradient; T detention time; n^O (12 mg/l) particle concentration in influent; floc concentration 1522 mg/l) (Parker *et al.*, 1971)

unless this detention time is provided, the required performance level cannot be achieved regardless of the velocity conditions. The latter may be regarded as representing the intensity of agitation. From the data given in Fig. 16 it can be seen that performance improves as detention time is prolonged. However, at low values of intensity of agitation, increases of intensity lead to more frequent collisions and improved performance, because of the greater influence of aggregation phenomena rather than breakup. As intensity increases beyond a critical point, breakup assumes a greater role and performance deteriorates with further increases in intensity.

There seems to be no data available on the effect of agitation on the flocculation of brewers' yeast probably because mechanical agitation and aeration is not a feature of the traditional batch process. However, the effect of the fluidization velocity on flocs of yeast in a tower fermenter has been discussed by Greenshields and Smith (1971). They observed that when superficial liquid velocities are greater than 0.14 cm^{-1}, which is the practical operating range, the floc size appeared to be little affected by an increase in liquid flow rate. However, the bed of yeast does expand as velocity increases in a manner similar to a normal fluidized bed, i.e. particles of mixed sizes are segregated, the smaller at the top of the bed. Owing to the decrease in the substrate concentrations of the liquid phase up the tower, natural decrease in floc size with increasing height is reinforced by the different growth rates. These factors counteract the effect of changes in the difference between floc and liquid densities with tower height. Greenshields and Smith (1971) also observed back-mixing and this affects liquid distribution and "solid" phase packing.

Clark and Lentz (1963) reported that, in submerged citric acid fermentations using pellets of *A. niger*, agitator speeds greater than 300 rpm brokeup more than 95% of the pellets within 24 hrs of inoculation regardless of the sparger used. It was observed that the pieces of mycelium (single filaments of hyphae, or filaments bearing one or more club-shaped branches) resulting from the crushing of pellets grew rapidly during the first 100 hrs of fermentation, but since these pieces were further fragmented as they increased in size the original size range (50–250 μm length) was maintained.

Blakebrough and Hamer (1963), studying the resistance to oxygen transfer, found that in the case of the pellet form of growth of *A. niger*, impeller speed had no adverse effect on the pellets in the range of impeller speed from 265 to 835 rpm. When the impeller was stationary the density of the pellets was such that they floated; when speeds greater than 835 rpm were used, the pellet structure was broken up. Similar behaviour was observed by Moscona (1961a) for embryonic cell aggregates (Fig. 17), the average diameter of the aggregates decreasing with increasing speed of rotation.

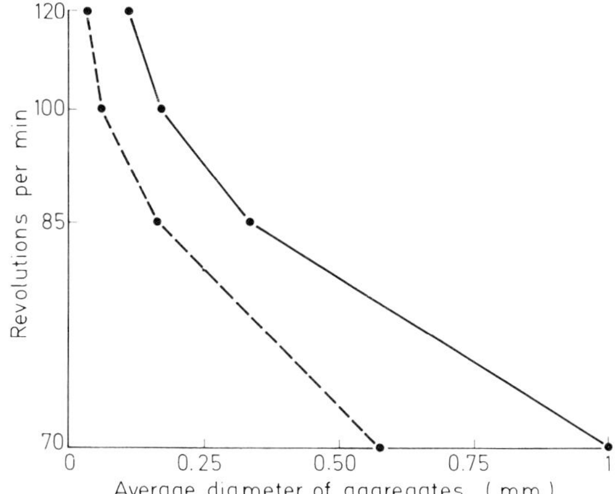

Fig. 17. Effect of agitation on size of embryonic cell aggregates. The broken curve represents chick-embryo retina cells; the solid curve, chick-embryo liver cells (Moscona, 1961a)

The effect of aeration in fermentation is two-fold; it causes agitation and satisfies the oxygen needs of the microorganisms. In an attempt to determine the effects of aeration rate on growth of *Morchella hortensis*, Litchfield *et al.* (1963) aerated culture vessels at rates of 0.25, 0.50 and 0.75 litre air/litre medium. min; this corresponded to sulfite oxidation values of 0.08, 0.15, and 0.20 mmolO_2/litre min. They observed that the mycelium produced at the 0.15 mmol aeration rate was somewhat filamentous as compared with the firm round pellets ranging from 0.5 to 0.8 cm in diameter obtained at the 0.08 mmol aeration rate. The mycelium produced at the 0.20 mmole aeration rate was filamentous, slimy and difficult to harvest.

Morris *et al.* (1973) claim that tower fermenters, when using their method of aeration and agitation, impose very little shear force on the medium or the organism present, thus facilitating control of the morphology.

The effect of dilution rate on the morphology of *A. niger* in tower fermenters has been investigated by Cocker and Greenshields (1975). At dilution rates < 0.1 hr^{-1} loose feathery colonies developed while at higher rates, the hyphae developed radially and evenly to form 1–5 mm spherical colonies with low hyphal packing densities. These changes were reversible, according to the dilution rate used.

The viscosity of the culture medium has also been shown to exhibit an influence on pellet size (Whitaker and Long, 1973). At low viscosities an inverse relationship was found between pellet diameter and culture viscosity when using *A. niger* and *P. chrysogenum*, while dispersed mycelial forms developed at higher viscosities.

Cocker and Greenshields (1975) found no appreciable differences in the pattern of morphological development of *A. niger* in tower fermenters of various sizes. However, Block *et al.* (1953) found that pellets of *Agaricus blazei* grew to an average diameter of 2.8 mm in shake flasks, whilst in aerated culture bottles they grew up to 25 mm diameter, but were not as compact.

3.4 Interfacial Phenomena

Renn (1956) proposed that, in the activated sludge process, the growing bacterial population concentrates at the air-water and water-solid interfaces. Both of these interfaces act as accumulators: they absorb and concentrate organic matter from solution, and they afford a favourable physical site for dense bacterial populations.

Stanley and Rose (1967) confirmed that gas bubbles provide a suitable interface at which the bacteria can collect (Fig. 18). They found that only a single layer of bacteria collected around each gas bubble, and substituting nitrogen or oxygen instead of air had no marked effect on clumping ability.

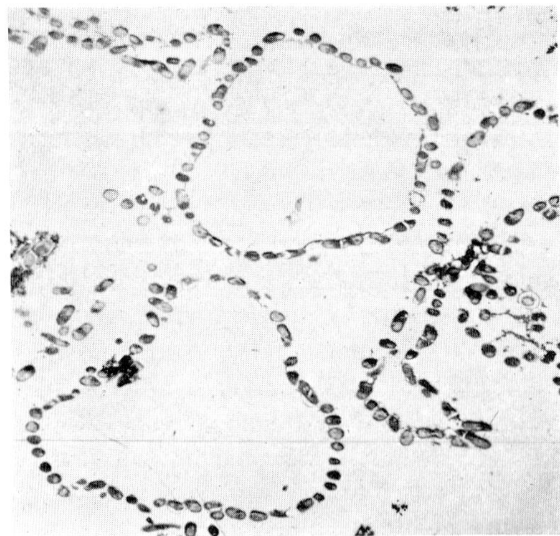

Fig. 18. Electron micrograph of a thin section through a clump of *Corynebacterium xerosis* NCIB 9956 formed on cooling a culture from 30 to 15° with fast stirring. × 5000 (Stanley and Rose, 1967)

The presence of an articulated network of filaments within activated sludge flocs has been shown by Parker *et al.* (1971). This network provides a "back-bone" for the building of particles by enmeshment and polymer bridge bonding. This supports belief in the importance of the water-solid interface in flocculation.

3.5 Chemical Factors

The presence of certain chemical elements and compounds in the medium appears to affect flocculation. These include both nutrients and metabolic products.

In brewing, the complexity of the fermentation medium necessitates the study of the effect of many substances. Rainbow (1966) reports that while fermentable sugars are known to prevent flocculation, or to cause the redispersal of flocculated yeast, there is a formidable list of other substances which have been reported to cause flocculation. These include furfural; a poly-saccharide material (treberin) isolated from six-rowed barley; proteins; colloidal wort components such as humic acid, melanoidins and phlobaphene; hard hop resin and ethanol.

Mill (1964b) has confirmed the finding by Eddy (1955) that certain specific sugars could disperse flocculent yeast. It has also been observed (Mill 1964a; Baker and Kirsop, 1972) that the presence of some nitrogeneous nutrients in the medium could delay the point in the growth cycle at which brewers' yeast becomes potentially flocculent. An example is of yeast completely dispersed at a urea concentration of 40%, but with the urea removed by washing with buffer, flocs reappeared. This effect of urea was also observed with *Schizosaccaromyces pombe* (Calleja, 1970).

In the case of fungal organisms it has been shown that the presence of trace metals, oils, ferrocyanide, and nitrogen affects pelleting. Clark *et al.* (1966) have reported that the addition of as little as 2 ppb of manganese to ferrocyanide-treated beet molasses, during citric acid fermentation by *A. niger* caused an undesirable change in the morphology of the organism from the normal pellet-like form to the filamentous form. Other metals tested did not visibly change pellet morphology. The effect of zinc concentration on the morphology of *A. niger* has been investigated by Cocker and Greenshields (1975), who found that growth was considerably inhibited at concentrations above 0.01% w/v while at 0.15% and above, densely packed pellets as large as 10–20 mm in diameter were produced.

The morphology of a mutant strain of *A. niger* was found by Millis *et al.* (1963) to be affected by the concentration of oil present in the medium. In the presence of 2% v/v oil the pellets were about 1 mm in diameter in a fermentation mash of very low viscosity. Whitaker and Long (1973) report that Baig *et al.* (1972) observed a similar phenomenon; addition of 2–10% oil caused *A. niger* to form smooth round pellets.

Steel *et al.* (1954) have found that ferrocyanide concentration is also important when using *A. niger,* as it influences both the type of growth and the rate of development of the inoculum; too low a level resulted in loss of the characteristic morphology, while too high a level retarded growth excessively. Clark (1962a) found that at concentrations of 10 ppm or less, large (up to 6 mm dia.) soft filamentous pellets were formed during the growth stage. At concentrations about 20 ppm round firm pellets were formed;

these developed to a diameter of 1–1.5 mm during the growth stage and to 3–3.5 mm during the first 3 days of the acid-producing stage. Little change in growth rate and ultimate pellet size was noted for ferrocyanide concentrations above 20 ppm.

The influence on the morphology of *A. niger* of the carbon/nitrogen ratio in the medium has been reported by Cocker and Greenshields (1975). At constant carbohydrate concentrations, a ratio of 9:1 militated towards spherical colonies with low hyphal packing density and diameters of 0.5–5 mm. At ratios around 15:1 (effectively nitrogen starvation) colonies up to 7 mm diameter developed while the hyphae were more highly branched and densely packed. At ratios higher than 15:1 an extreme form was observed in which radial organisation was not apparent and the branching-frequency was extremely high.

The presence of enzymes in the medium and their effect on flocculation has been investigated. Stanley and Rose (1967) observed that while hyaluronidase, lipase, lysozyme chloride or phospholipase had no measurable effect on the ability of *C. xerosis* to clump, proteolytic enzymes had some effect. Papain completely removed the ability to form clumps and trypsin led to the disappearance of much of the adhesive material that surrounded shadowed preparations of bacteria on electron micrographs. Eddy and Rudin (1958) observed that treatment of isolated yeast cell walls with papain destroyed their flocculence, with simultaneous solubilisation of a protein-mannan complex. Calleja (1970) reported that the deflocculation of *S. pombe* by trypsin or papain was irreversible and that the effect of neither reagent could be removed by washing.

The above phenomenon is exploited in studies on the aggregation of embryonic cells. These cells are trypsin-dissociated and then the effect of various factors on reaggregation is investigated. The effect of colloids and chelating agents on flocculation has also received attention. In his treatment of biological flocculation McKinney (1956) reports that the addition of colloids has a profound effect on bacterial floc formation. With the exception of certain industrial wastes, inorganic and organic colloids are usually present in wastes. The inorganic colloids include various soil colloids, metallic oxides, metallic phosphates and some metallic carbonates. The organic colloids are primarily lipo-proteins with some polysaccharides also present. Inorganic colloids, like organic colloids, are primarily negatively charged. Studies with calcium phosphate colloids showed that bacteria were readily adsorbed to the inorganic aggregate when they lacked sufficient energy to pull away. Two factors are involved in the flocculation of bacteria and organic colloids; namely the electrical charge and the biological utilization of the colloid. The organic colloids and the bacteria do not have sufficient charge to prevent flocculation. If the bacteria are not able to utilize the colloid as food, then the organic colloid acts as an inorganic colloid. If the bacteria are able to utilize the colloid as food, the colloid has little effect on initial floc formation. Rapid degradation yields considerable energy and momentarily prevents flocculation, while slow metabolism might not yield sufficient energy to prevent flocculation.

Taylor and Orton (1973) investigated the inhibitory effect on the flocculation of *S. cerevisiae* of anions which complex calcium, e.g. EDTA and potassium salts of phosphate, fluoride, bicarbonate, oxalate, citrate, diglycollate and nitrilotriacetate. They found the lowest inhibiting concentration for each salt to be roughly in proportion to its complexing power. EDTA was particularly effective as it inhibited flocculation of the three strains tested. The inhibitory effect of EDTA was reversible and cells flocculated after resuspension.

It has also been reported by Choudhary and Pirt (1965) that the morphology of *A. niger* in shake flasks is highly sensitive to the presence of metal-complexing agents: e.g. EDTA and diaminocyclohexane. Specifically EDTA and ferrocyanide induced the formation of small (1.3 mm dia.), smooth, granular, easily dispersed pellets of mycelium, which contrasts with the viscous conglomerate of larger pellets (2.0 dia.) with filamentous peripheries formed in the absence of chelating agents.

3.6 Biological Factors

The initial microbial concentration (inoculum level), the presence of other microorganisms or strains, and floc concentration, have been mentioned as exerting an influence on flocculation.

As a result of their work on activated sludge floc, Parker *et al.* (1971) claim that the floc concentration and the concentration of suspended particles, whether biological or otherwise, affect floc breakup and aggregation.

The effect of spore level on the morphology of *A. niger* has been investigated by Steel *et al.* (1954). When the spore level was increased the pellet growth tended to change from the retarded type showing scant growth towards the standard pellet type. In the latter the individual pellets were 0.2 to 0.5 mm diameter with precipitated matter enmeshed in the centre. The lateral hyphae were short, thickened, vacuolated, and granular with short club-like branches. Foster (1949) suggests that large spherical pellet-type growth is to be avoided when studying the metabolic activities of fungi. He regarded the most common fault in the submerged cultivation of fungi to be the use of too small an inoculum. When the number of viable cells becomes too small, the desired homogeneous suspended type of growth consisting of small individual colonies or clumps of mycelia fails to develop. Instead, the relatively few viable cells each develop into large individual colonies that are usually spherical and range from the size of small peas to marbles or even larger.

Attention has been given to the effect of the presence of many microbial species and strains in the system of flocculation. In biological waste water treatment pure-culture bacterial flocs do not exist normally. In discussing this, McKinney (1956) reports that the bacteria exist in association with certain free swimming ciliates and stalked peritiches. He examined a common free-swimming ciliate, *Tetrahymena gelii*, alone and in association with various bacteria. *T. gelii* formed flocs in both synthetic and sterile domestic sewage. As the protozoa reached the declining growth phase, large spherical bodies were noted within the cells. Following death and lysis, these spherical bodies were liberated into the liquid environment and readily flocculated. In the presence of bacteria, floc built up as a combination of bacteria and the spherical bodies from the protozoa. As the floc increased in size and age, both the bacteria and the spherical bodies from the protozoa lost their individuality and became a part of the amorphous floc.

In relation to the employment of complex mixtures of various strains of yeast Eddy (1958) reported the phenomena of mutual aggregation. He found that various strains of brewers' yeast which are non-flocculating precipitate readily in the presence of an appropriate partner, i.e. another strain. A series of strains of *Saccharyomyces cerevisiae* grown in malt-extract medium were tested two at a time, using every possible combination. It was observed that, though the interactions were relatively specific, they were not confined merely to a few exceptional pairs of yeasts. A further survey was carried out where a series of strains of *Saccharomyces carlsbergensis* were tested two at a time in every combination, both among themselves and with strains from the previous *S. cerevisiae* series. Mutual aggregation was also observed with a large number of combinations.

4. Characteristics of Microbial Flocs

The description and characterisation of the biological flocs present in a biochemical reactor is as essential in determining the rate of substrate uptake and product formation as knowing the concentration of the limiting nutrient. Unfortunately the deformable nature of biological flocs prevents the establishment of satisfactory measurement procedures.

Important properties of flocs include morphology, size range, number of viable and non-viable cells per unit volume, density, dry weight and water content. Both the micro- and macro-morphology, i.e. internal and external morphology, of a floc have to be considered.

4.1 Properties

4.1.1 Morphology

Stanley and Rose (1967) examined clumps of *C. xerosis* microscopically. Electron micrographs showed the bacteria to be connected by adhesive material which did not appear to be localized but to be distributed over the bacterial surface. The surfaces of the bacteria were covered by 'wart-like' lumps. A thin section of a clump showed it to be hollow and to have been formed by a single layer of bacteria collecting round a gas bubble. The ability of the bacteria to remain at least partly in contact, after preparation of the material for electron microscopy, testified to the mechanical strength of the adhesive material.

The presence of certain microorganisms in a mixed microbial floc adds to its strength. Parker *et al.* (1971) suggested the presence of filaments within the floc, (Fig. 19) while McKinney (1956) observed that dead protozoa liberated solid bodies on which bacteria adhered forming flocs.

Different strains of brewers' yeast may, under normal working conditions, be flocculent or non-flocculent. Greenshields and Smith (1971) found that the different brewers' yeast used in their tower fermenters may be classified as follows:

a) non flocculent, i.e. those which did not attain more than 3% w/w concentration and were rapidly washed out from the tower;

b) flocculent-physically limited; these attained 20–30% w/w and formed fine flocs up to 0.2 cm in diameter. At any particular wort-sugar concentration there was a critical

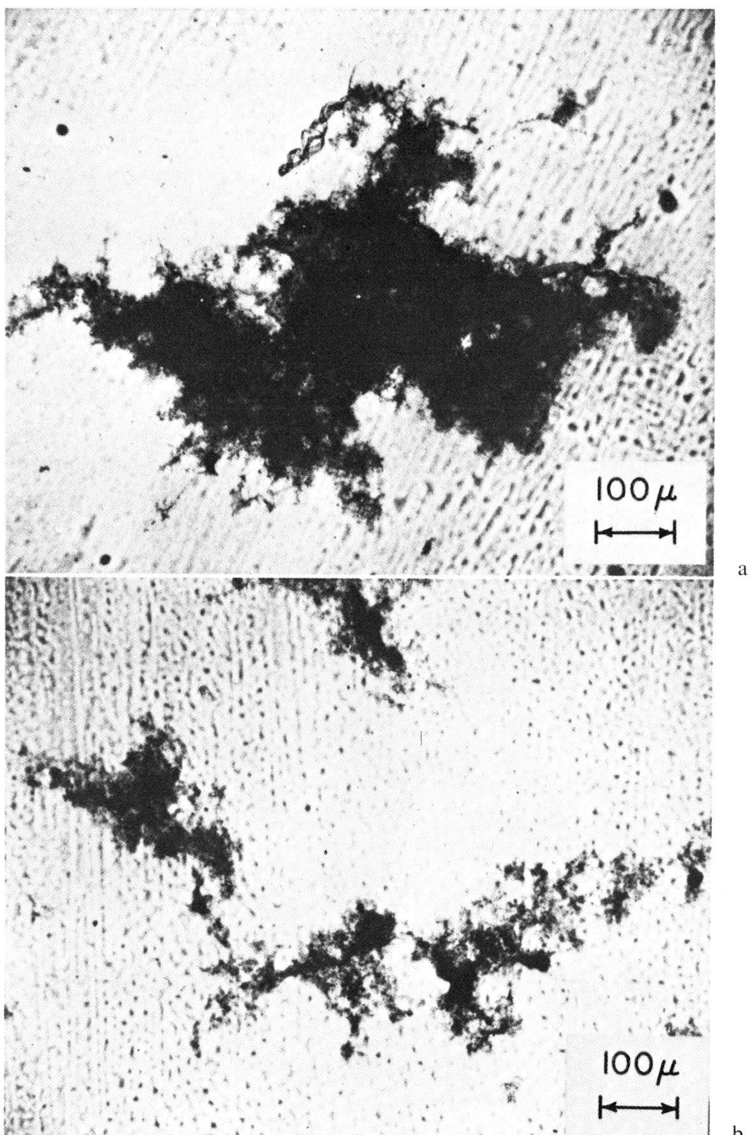

Fig. 19 a and b. Typical activated sludge flocs (a) low shear (b) high shear (Parker *et al.*, 1971)

"space velocity"; i.e. volumetric flow rate of wort/void volume of tower, which if exceeded caused complete wash-out of this type of yeast;
c) flocculent-fermentation limited; these attained concentrations of 25–40% w/w having a heavy "sticky" floc up to 1.3 cm in diameter. They were retained in a tower up to relatively high space velocities. However, before the critical space velocity was reached

for these yeasts the sugar conversion normally associated with beer production could no longer be achieved.

Electron micrographs, of 4-days old colonies of *S. cerevisiae* have been presented by Passmore (1973). Comparison between glutaraldehyde-fixed and unfixed colonies showed that the cells of the unfixed specimen appeared coated with a slimy material; glutaraldehyde treatment tended to remove this slime layer and gave a clearer micrograph (Fig. 20).

A photograph of a typical yeast floc presented by Smith and Greenshields (1974) indicates incomplete separation after budding (Fig. 21).

There appears to be considerable variation in the morphology of fungal pellets as growth proceeds. Some species have pellets with a smooth exterior whilst others are hairy; some are compact, whilst others are loosely held together. Martin and Waters (1952) have observed the morphological characteristics of *A. niger* pellets. During the period 12–18 hrs after inoculation, the small pellets (about 0.75–1.0 mm dia.) began to aggregate into loose flocculent masses, and the suspension began to clear. After 18 hrs the aggregates broke up into individual hard, smooth pellets. At this stage, if the air flow was stopped the pellets tended to rise, while the same action after 30 hrs resulted in the pellets tending to settle out. Washed pellets taken after 30 hrs or later were creamy white, smooth, gravelly to the touch, and of mean size in the region 1.0–2.0 mm diameter. Steel *et al.* (1954) have observed a complete range of morphological types of *A. niger* growth (Fig. 22). The extremes of this range are represented by the filamentous type, showing abundant growth, and the retarded type showing scant growth with poor fermentation results. The pellet type was an intermediate between the two extremes and was selected as the "standard" because it gave good results in fermentations for citric acid production. The individual pellets were 0.2–0.5 mm diameter with dark centres that formed at about 16–18 hrs. The "standard" pellet had a somewhat retarded appearance in the early stages of growth but rapidly developed lateral growth.

The internal structure of fungal pellets have been shown to vary a great deal. Martin and Waters' (1952) microscopic examinations of *A. niger* pellets have shown that the pellets are dense, spherical with limited lateral hyphae, and with much precipitate enmeshed in the pellet. The individual hyphae are short, greatly thickened, vacuolated and granular with short club-like branches. Particles of precipitate adhering to the hyphae were also observed. Steel *et al.* (1954) reported similar findings.

A recent morphological study of *A. niger* has been carried out by Cocker and Greenshields (1975) leading to the comprehensive classification of pellet types illustrated in Fig. 23.

Pirt and Callow (1959) found that pellet formation of *Penicillium chrysogenum* was linked with the production of an aberrant form of mycelium. The characteristic of this aberrant form was the presence of short, much branched, swollen and often distorted hyphae. The normal filamentous type was characterised by long, thin hyphae of constant thickness and few branches. They observed the initiation of pellet formation to be aggregation of the hyphae within the individual fragments of mycelium.

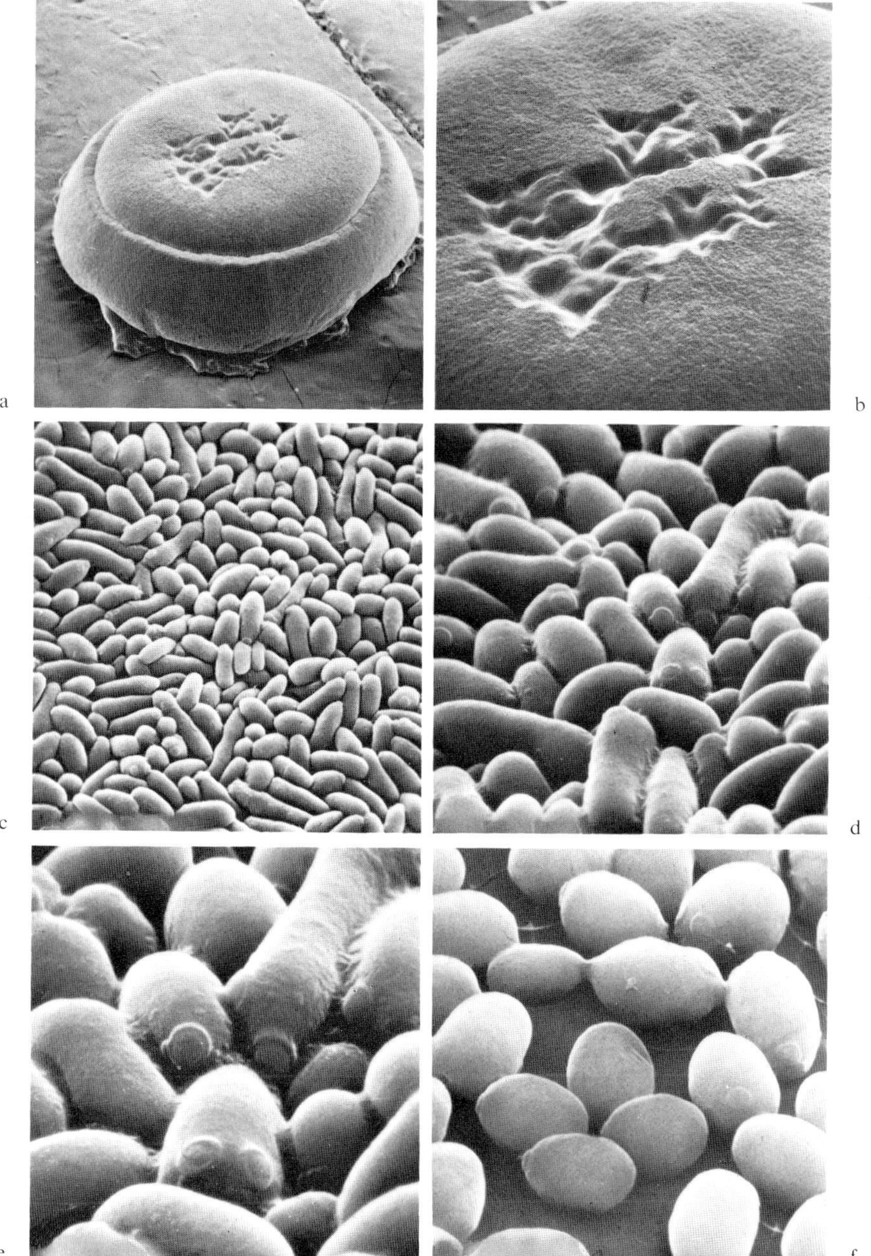

Fig. 20 a–f. Micromorphology of yeast (*Saccharomyces uvarum*). (a) Typical colony (× 22); (b) Detail of top of colony (× 63); (c) Cells with prominent bud and birth scars (× 1 200); (d) Cells with prominent bud and birth scars (× 3 000); (e) Cells with prominent bud and birth scars (× 6 000); (f) Fixed, discrete cells on glass (× 3 400) (Passmore, 1973)

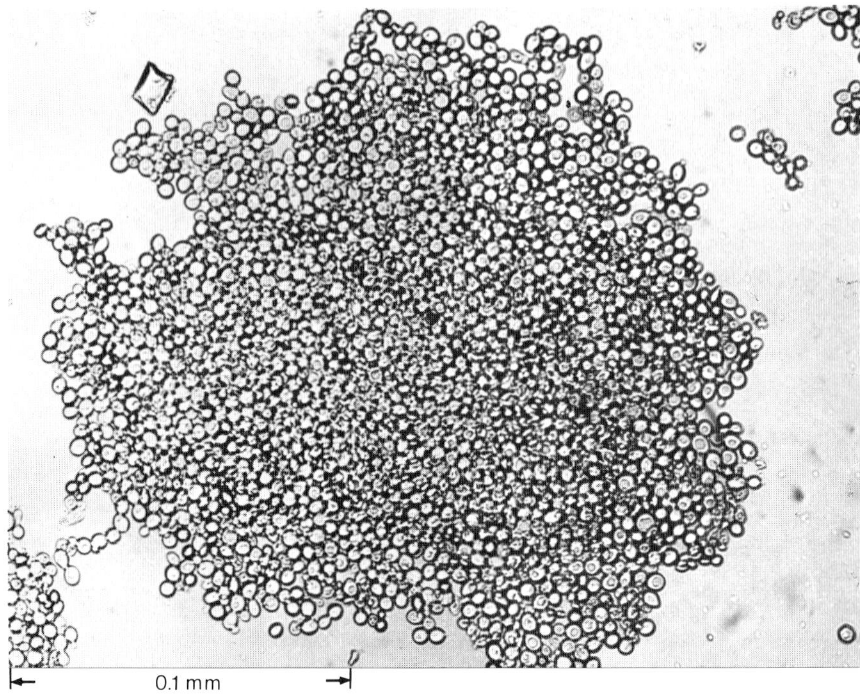

Fig. 21. Photograph of yeast floc taken from a tower fermenter (Smith and Greenshields, 1974)

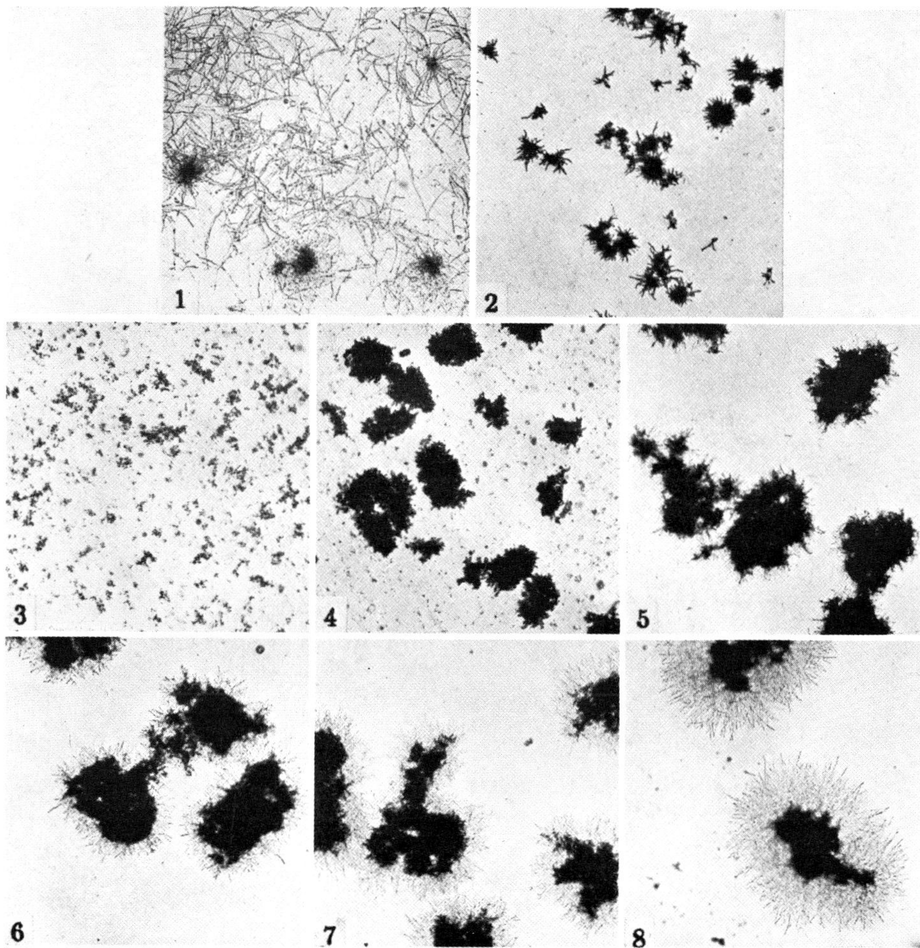

Fig. 22. Morphological types of *A. niger* (Steel *et al.*, 1954). 1–Filamentous type; 2–Retarded type; 3–8–Development stages of standard pellet-type; 3–Immature stage (14 hrs); 4–Early prestandard stage (16 hrs); 5–Middle prestandard stage (18 hrs); 6–Late prestandard stage (20 hrs); 7– Standard stage (22 hrs); 8–Poststandard stage (25 hrs)

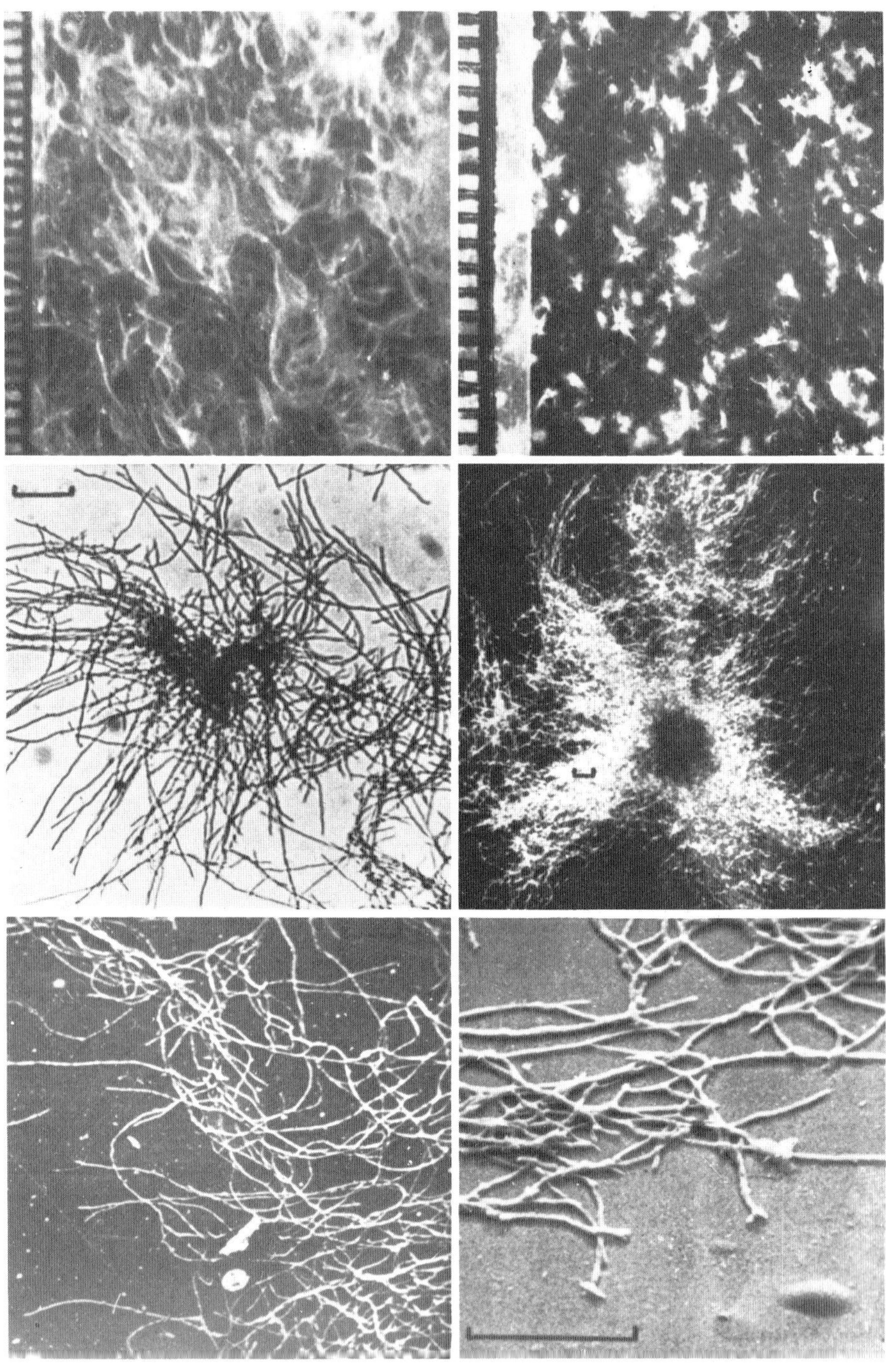

Type 1 Type 2

Fig. 23 a

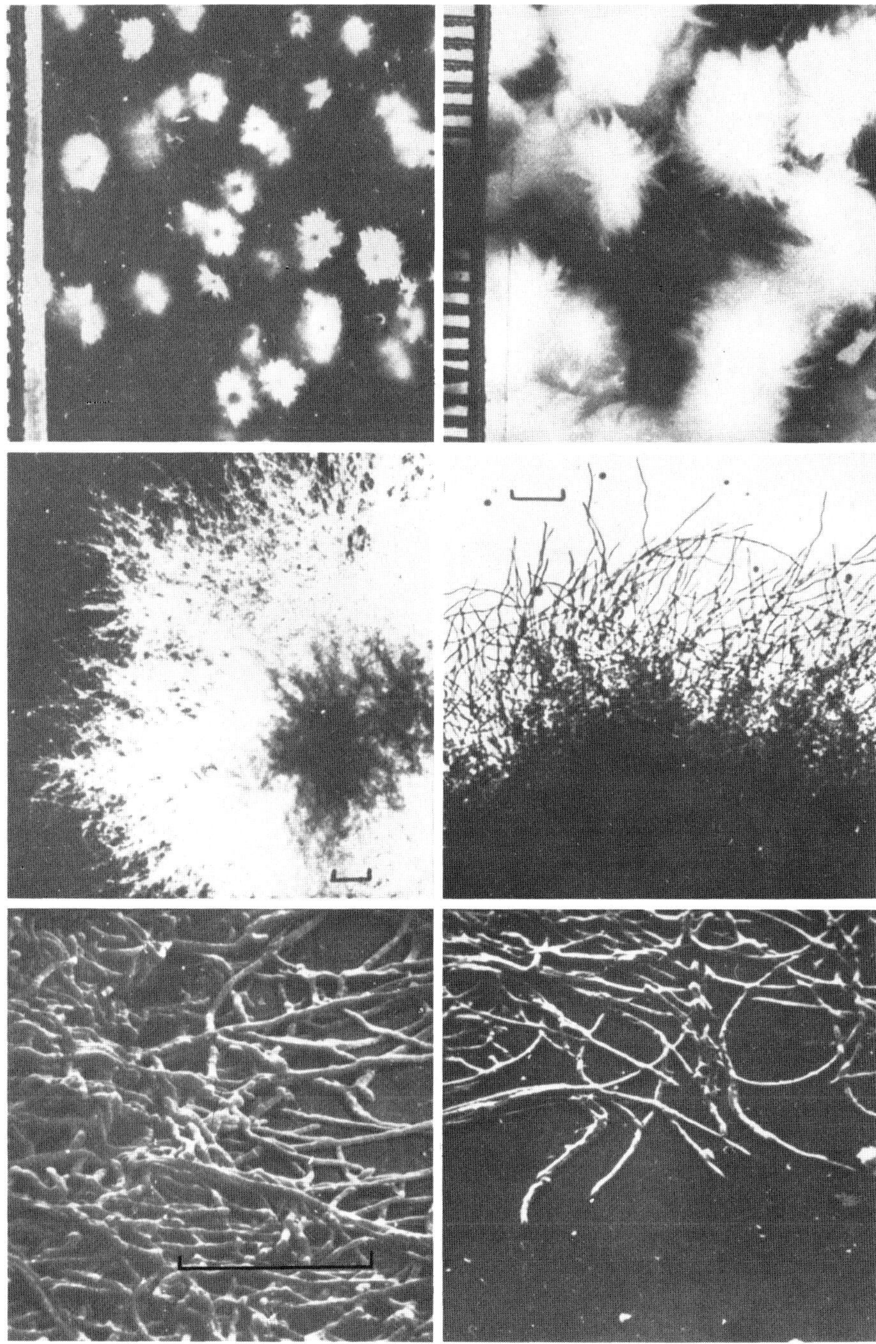

Type 3 Type 4

Fig. 23 b

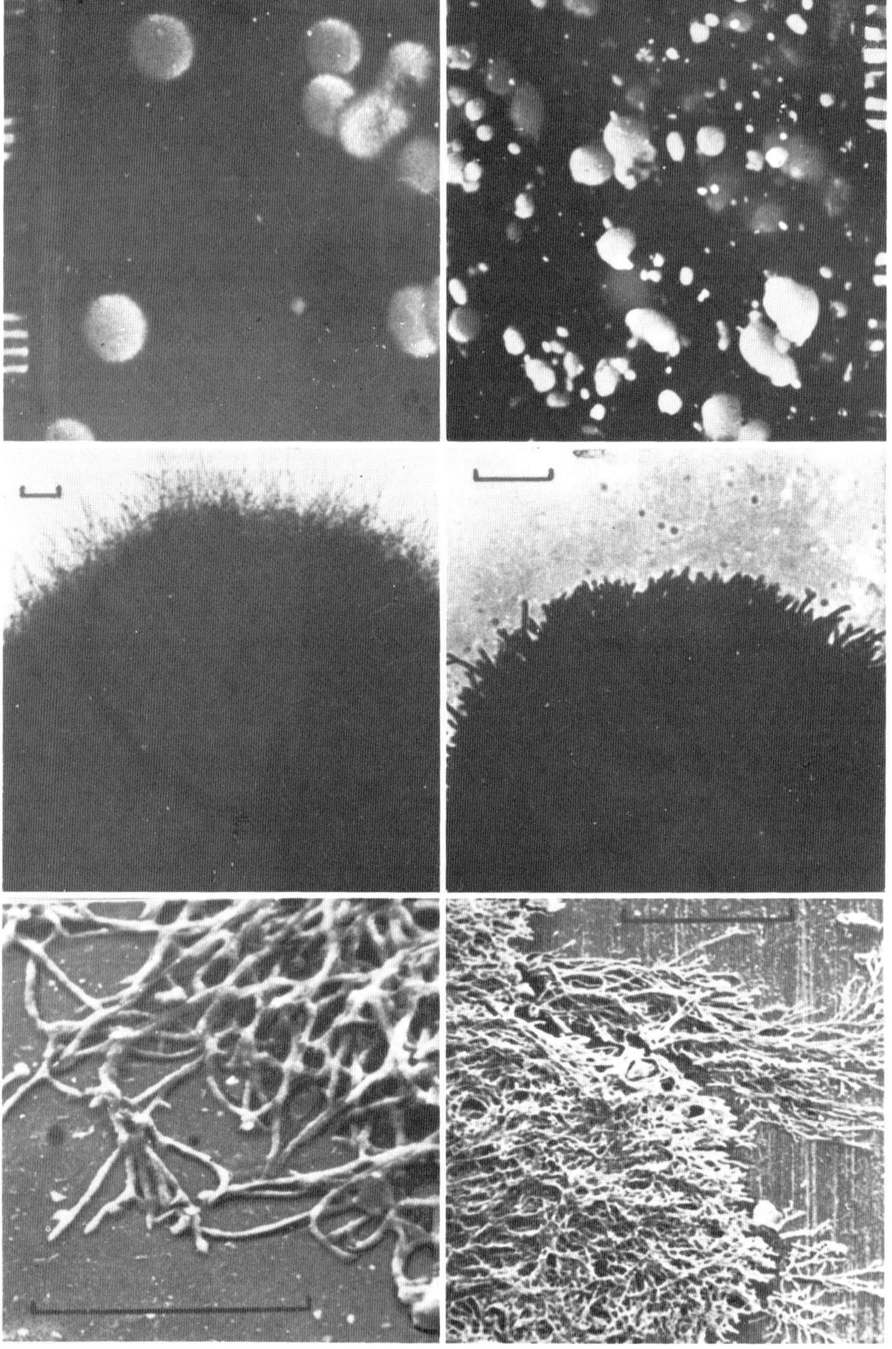

Type 5 Type 6

Fig. 23 c

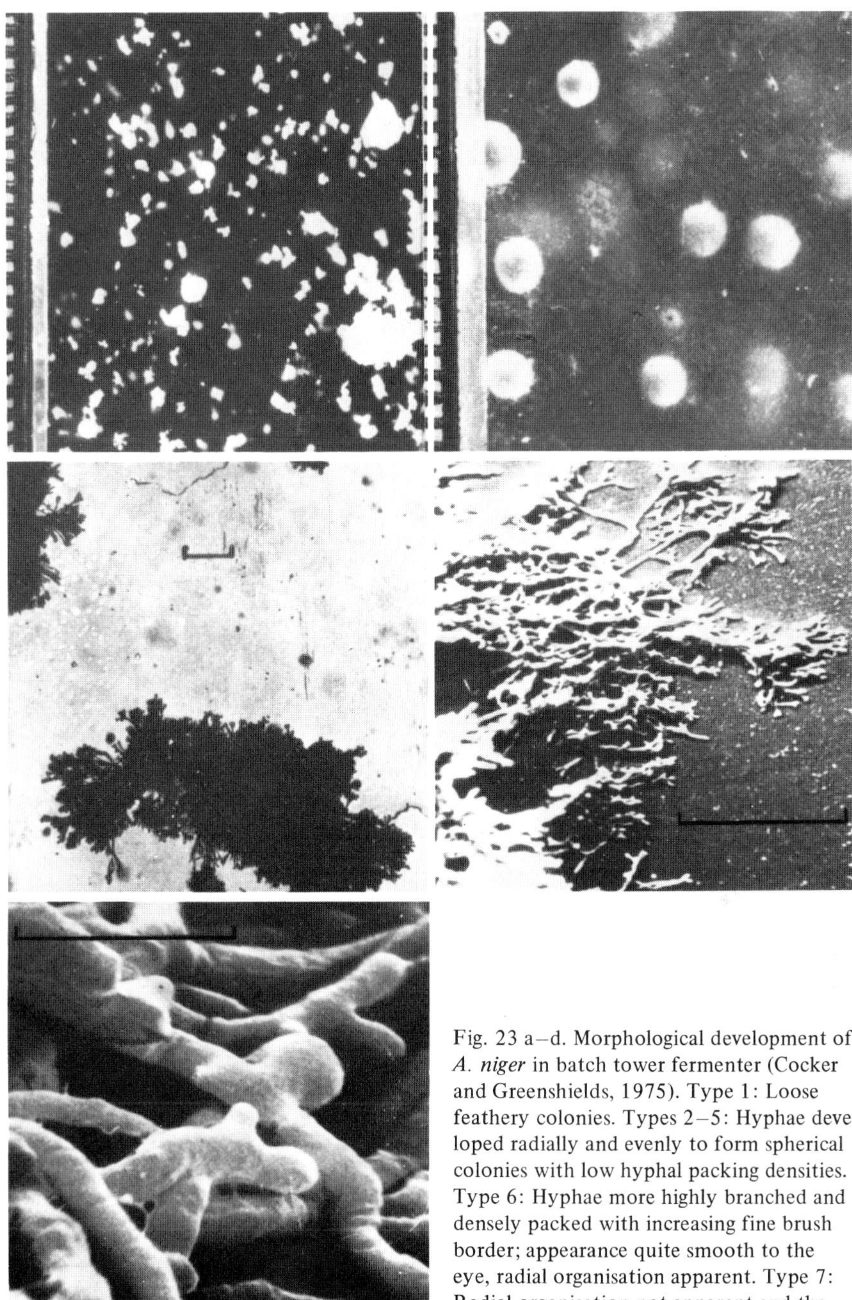

Fig. 23 a–d. Morphological development of *A. niger* in batch tower fermenter (Cocker and Greenshields, 1975). Type 1: Loose feathery colonies. Types 2–5: Hyphae developed radially and evenly to form spherical colonies with low hyphal packing densities. Type 6: Hyphae more highly branched and densely packed with increasing fine brush border; appearance quite smooth to the eye, radial organisation apparent. Type 7: Radial organisation not apparent and the branching frequency is extremely high.

Type 7

Fig. 23 d

4.1.2 Size Ranges

It is perhaps worthwhile ro recall microorganism sizes as a prelude to discussing floc sizes. Hawker *et al.* (1963) give a table of shape and size of representative microorganisms. Spherical, rod shaped and filamentous bacteria have sizes of 1.0 μm, 0.7–2.8 μm x 0.7–1.5 μm and 0.5–1.2 μm diameter respectively. Filamentous fungi have diameters between 11–42 μm while the ellipsoidal fungus *S. carlsbergensis* has a size 5–10.5 μm x 4–8 μm.

Table 7 gives the size range of various types of flocs. Unfortunately many workers fail to mention floc size and those that do usually fail to give even an approximate indication of size range. Indeed little information on floc size distribution is available in the literature. An exception is the work of Parker *et al.* (1971) in connection with the physical conditioning of the activated sludge floc. These workers carried out a floc-size distribution study in which they took the maximum floc dimension as the criterion of size. The data given in Fig. 24 are typical; the distribution is binodal, and very few particles occurred in the range 5 to 25 μm. Floc break-up occurs as a result of filament fracture and surface erosion and this probably explains the presence of two maxima. The floc size distribution is typified by a very narrow band of primary particles of 0.5–5 μm in size, i.e. unflocculated suspended solids, and a very wide band of particles from 20 μm up to 5 000 μm classed as flocs.

Table 7. Characteristic size of flocs

Microorganism	Floc size range	Reference
Agaricus blazei (mushroom)	2.8–25 mm dia.	Block *et al.* (1953)
A. niger	0.2–0.5 mm dia.	Steel *et al.* (1954)
A. niger	0.75–2.0 mm dia.	Martin and Waters (1952)
Brewers' Yeast		
a) physically limited	up to 0.2 cm dia.	Greenshields and Smith (1971)
b) fermentation limited	up to 1.3 cm dia.	Greenshields and Smith (1971)
Mixed Bacterial Cultures	25–5 000 μm	Parker *et al.* (1971)

4.1.3 Viable/Non-Viable Cell Ratio

In the activated sludge process, bacteria, protozoa, inorganic and organic colloids, all contribute to the formation of flocs. The recycle of activated sludge floc controls the viable/non-viable cell ratio in the floc for any given organic substrate concentration. Low recycle results in a very active floc with a small non-viable fraction. The non-viable fraction has a significant absorptive capacity for the organic substrate though rather less than the viable fraction. Thus for a given aeration period and a given waste, flocs of high viability produce an effluent containing less organic matter, but have a higher oxygen demand than flocs of low viability (McKinney, 1956).

Clark (1962b) has studied the internal structure of *A. niger* pellets. Under optimum fermentation conditions, each pellet developed as a round mass of mycelium of uniform consistency during the first 24 hrs of aeration; subsequently a dense crust of growth

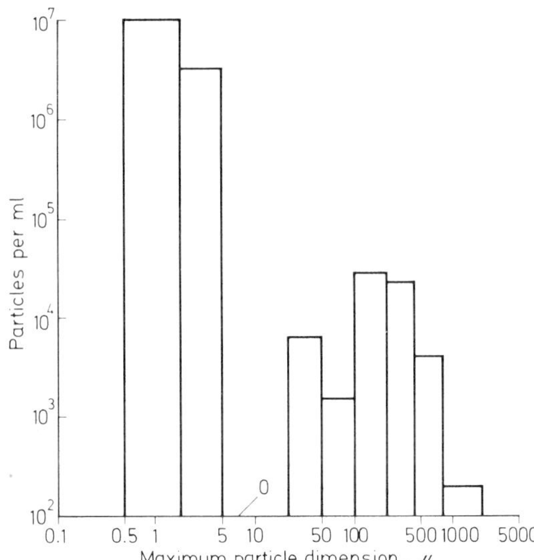

Fig. 24. Size distribution of activated sludge floc in a stirred tank (High Shear) (Parker *et al.*, 1971)

formed at the periphery of the pellet and autolysis of cells began at the centre. At the end of the fermentation (140 hrs), the pellets consisted of a shell of mycelium occupying less than 50% of the pellet volume. Autolysis was thought to result from the resistance to the passage of nutrients and oxygen to cells inside the pellet, by the dense growth at the surface.

4.1.4 Density and Water Content

In connection with hydrodynamic studies in tower fermenters, James and Smith (1975) examined flocs of *S. cerevisiae* and pellets of *A. niger.* They found that the average dry density of *S. cerevisiae* was 1.32 gml^{-1}, while that of *A. niger* varied significantly during the course of a batch fermentation; 1.24 gml^{-1} was obtained for mature mycelium. After evaluating wet cell density, they suggested a value of about 72% by volume for the liquid content of single cells.

James and Smith (1975) also estimated aggregate densities by measuring the terminal velocities of microbial aggregates of known size. The volume fraction of liquid in the aggregate was then computed from a knowledge of dry density. For many pellet morphologies the volume fraction of liquid was as high as 99%. In the case of yeast flocs the corresponding values were in the range of 90–95%. It is of interest to note that Green *et al.* (1965) reported that the average percentage dry matter in microbial film grown, at various temperatures between 5–30° C, on settled sewage was 3–7%.

4.2 Measurement of Flocculation and Floc Size

Numerous techniques have been used for measurement of flocculation and floc size including: measurement of optical density before and after flocculation and gravity sedimentation; measurement of the time required to filter a given amount of microbial suspension at constant pressure; membrane filters; and microscopic examination. Light-scattering techniques are widely used and are regarded as the most convenient and reproducible. Stanley and Rose (1967) measured the extent of clumping of *C. xerosis.* They placed 5 ml of the suspension into a glass cell (1 cm light path) and then allowed a period of 5 min during which clumps of bacteria rose to the surface. By measuring the turbidity of the underlying fluid, and comparing this with the turbidity of the suspension before clumping occurred, they estimated the percentage dry weight of bacteria that had clumped. This procedure could be regarded as a measure of the flotation rate of flocs. A similar method was used by Pavoni *et al.* (1972), who measured flocculation as percentage transmission at 690 μm following settling.

A more precise procedure using the sedimentation method was followed by Mill (1964a). Potential flocculence was determined with yeast cells suspended in a 0.5 M sodium acetate buffer (pH 4.6) containing 0.1% $CaCl_2$. The gross rate at which the cells sedimented in this buffer was taken to be exclusively dependent on the size of the flocs. The suspension was allowed to stand in a glass cell in a colorimeter and readings of turbidity were taken at timed intervals and converted to cell dry-weight. The steepest slope of the cell dry weight *versus* time curve was taken as a measure of the sedimentation rate. The logarithm of this rate was considered to provide a good measure of the relative flocculence of the cells in the buffer, and hence of the potential flocculence of the yeast cells in the culture under consideration.

Direct measurement of the floc sizes was carried out by Parker *et al.* (1971) in their studies of the physical conditioning of activated sludge floc. The procedure they followed involves five steps:

a) sludge withdrawal and dilution under *in situ* turbulent conditions;
b) sampling with wide bore (8000 μm) pipette;
c) particle impingement on a membrane filter;
d) drying and particle staining with methylene blue solution followed by mounting on a slide and
e) particle counting.

Other workers mention using cake filtration time (Pavoni *et al.*, 1972) and microscopic observation (Calleja, 1970). Kobayashi *et al.* (1973) measured the diameters of the *A. niger* mycelial pellets by microscope; an average of about ten pellets was taken as a representative value.

5. Overall Rate of Substrate Uptake

The previous sections have largely been concerned with the degree to which microorganisms flocculate under a variety of environmental conditions. Fermenter design (Atkinson, 1974) requires knowledge of the overall rate of substrate uptake by individual flocs and is not so much concerned with the mechanism of flocculation but more with the fact of flocculation. Nevertheless it is to be expected that the overall specific rate of substrate uptake is likely to depend upon the microstructure of the flocs and this, in turn, depends upon the environment in which the flocs have developed (Section 4).

Atkinson and Daoud (1968) suggested the diagramatic representation shown in Fig. 25, for microorganisms arranged in the form of flocs. In this model the microorganisms are assumed to be distributed uniformly throughout a biochemically inert intercellular gel.

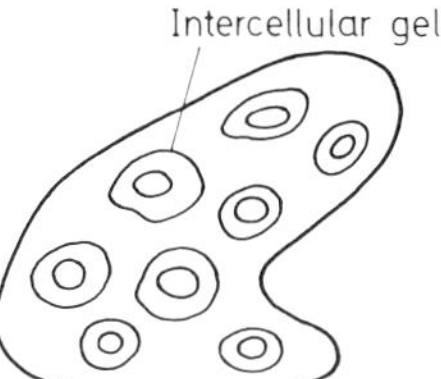

Fig. 25. A model for a microbial floc

The volume fraction of this region has been variously reported for centrifuged microorganisms as 23–38%, though the variation is easily attributable to the different experimental procedures adopted (Conway and Downey, 1950; Bettlestone, 1930). Since the microbial mass retains its cellular fluid (mainly water) when normally employed, it might be expected that these estimates of the "free space" are, if anything, lower than those pertaining in practice. It is of some interest to compare microbial mass with a random bed of packings where similar values of the "free space" occur (Coulson and Richardson, 1964, 1968).

The physical model for microbial mass, i.e. microorganisms dispersed in a biochemically inert gel, has considerable similarity to that for a porous catalyst.[1] Both systems consist of interconnected channels through which the reactants must diffuse before reaching the "active" surface. Indeed Atkinson and Daoud (1968) emphasised the analogy between microbiological reations and heterogeneous catalysis.

The kinetic description of an individual microorganism, which results from a detailed physico-biochemical analysis, is so complex that if complete cognisance of all its features had to be taken into account to achieve a mathematical model, the problem would be beyond comprehension.

This is illustrated by the work of Perret (1960) in a conceptual discussion of relatively simple models for the metabolic pathways within a cell. Fortunately, experience of chemical kinetics (Petersen, 1965) indicates that such an interpretation would be unnecessarily complicated and that a model based upon considerable simplifications could lead to a meaningful result.

The growth rates of freely suspended microorganisms have been shown to depend upon the concentrations of the nutrient medium's constituents. These include not only the basic carbon and energy source, but also such limiting factors as amino-acids, nitrogen sources, inorganic salts, and oxygen (Novick, 1955). The algebraic description of this

[1] Highly porous particles with large internal surface area (say 200 m^2/g) coated with catalyst are used extensively in the chemical industriy (Thomas and Thomas, 1967).

dependence is most commonly expressed by the empirical Monod equation, (Monod, 1949; Cohen and Monod, 1957):

$$G = \frac{G_{max} C^*}{K_M + C^*} \tag{1}$$

where G = the specific growth rate
G_{max} = the maximum specific growth rate
K_M = a system coefficient
C^* = the substrate concentration in the aqueous solution adjacent to the microorganism.

Equation (1) is a two-parameter equation and provides a relationship between the *local* specific growth rate (G) and the *local* substrate concentration at any position within a floc. Studies associated with the structure and biochemistry of microorganisms have provided some insight into the empirical parameters G_{max} and K_M (Cohen and Monod, 1957, Atkinson, 1974).
An equation, with similar properties to Eq. (1), suggested by Tessier (1942) has failed to gain equal recognition, probably because the Monod equation is algebraically identical to the Michaelis-Menten equation of enzyme catalysis.
Further, three-parameter equations by Moser (1958), Contois (1959) and Powell (1967) have attempted to account for deviations from the Monod equation (Table 8). Powell (1967) suggested that such deviations were due to diffusional limitations within the membrane layers of the cell, although it is unclear as to whether any degree of flocculation occured during the experiments upon which this conclusion was based. At this stage it is difficult to justify the use of any equation other than the Monod equation in the development of equations intended to describe the overall rate of substrate uptake by a microbial floc. For a simple growth-associated system Eq. (1) becomes:

$$\frac{R}{R_{max}} = \frac{C^*}{K_M + C^*} \tag{2}$$

When a diffusional limitation exists the substrate uptake by the floc will be reduced because the local substrate concentrations within the floc will be less than that adjacent to the surface of the floc. Thus it is appropriate to introduce an effectiveness factor (λ):

$$\frac{R}{R_{max}} = \lambda \frac{C^*}{K_M + C^*} \tag{3}$$

where $0 < \lambda \leqslant 1$.

5.1 Biological Rate Equation (BRE)

An algebraic expression relating the effectiveness factor of Eq. (3) to the external substrate concentration (C^*), the characteristic size of the microbial mass, the effective diffusion coefficient of the substrate (D_e), and the biological parameters G_{max} and K_M, has been provided by Atkinson and Davies (1974) for the analogous problem of the

Table 8. Suggested forms of rate equation

Author	Equation	Basis	No. of biological parameters	No. of micro-organisms in floc
Tessier (1942)	$G = G_{max}(1 - e^{-C^*/B_1})$	empirical	2	1
Monod (1949)	$G = G_{max}\frac{C^*}{K_M + C^*}$	empirical	2	1
Moser (1958)	$G = G_{max}\frac{(C^*/K_M)^{B_3}}{1 + (C^*/K_M)^{B_3}}$	empirical	3	1
Contois (1959)	$G = G_{max}\frac{C^*}{K_M g(M) + C^*}$	empirical	3	1
Mueller *et al.* (1966)	$R = R_{max}(1 - x_c^3)$ where $0 \leqslant x_c \leqslant 1$; $\frac{2\,C^*/K_M}{3(k_2 V_p/A_p)^2} = 1 - 3x_c^2 + 2x_c^3$	diffusion with biochemical reaction (zero order)	3	any number
Powell (1967)	$G = \frac{G_{max}}{2(k_2')^2}[(1 + (k_2')^2 + k_3 C^*) - \{(1 + (k_2')^2 - k_3 C^*)^2 + 4k_3 C^*\}^{1/2}]$	diffusion with biochemical reaction (Michaelis-Menten)	3	1
Kobayashi *et al.* (1973)	$R = (\frac{\lambda_0 + \lambda_1 K_M/C^*}{1 + K_M/C^*}) R_M \frac{C^*}{K_M + C^*}$ where $\lambda_0 = \begin{cases} 1 \text{ when } 0 < m' \leqslant \sqrt{3} \\ 1 - (\frac{1}{2} + \cos\frac{\psi + 4\pi}{3})^3 \text{ when } m' > \sqrt{3} \end{cases}$ $\lambda_1 = \frac{3}{m'}(\frac{1}{\tanh m'} - \frac{1}{m'})$ $m' = \frac{3}{\sqrt{2}} k_2 \frac{V_p}{A_p}(\frac{K_M}{C^*})^{1/2}$ $\psi = \cos^{-1}(\frac{6}{(m')^2} - 1)$	diffusion with biochemical reaction (Michaelis-Menten)	3	any number

overall rate of substrate uptake by microbial films. The overall substrate uptake rates by flocs and films are related by

$$\frac{R}{R_{max}} = \frac{N}{N_{max}} \tag{4}$$

where N is the rate of substrate uptake per unit external area of microbial mass. It follows that the equation of Atkinson and Davies (1974) can be used to describe the overall rate of substrate uptake by microbial flocs, provided diffusion with reaction in a "slab" geometry can be equated with that in a floc geometry.

Figure 26 provides a comparison between the analytical solutions for diffusion with first order reaction in slab, cylinder and sphere geometries. This figure suggests that if the characteristic size is interpreted as the ratio of the particle volume to the external area, then virtually identical numerical values of the diffusional parameter are obtained for a given value of the effectiveness factor.

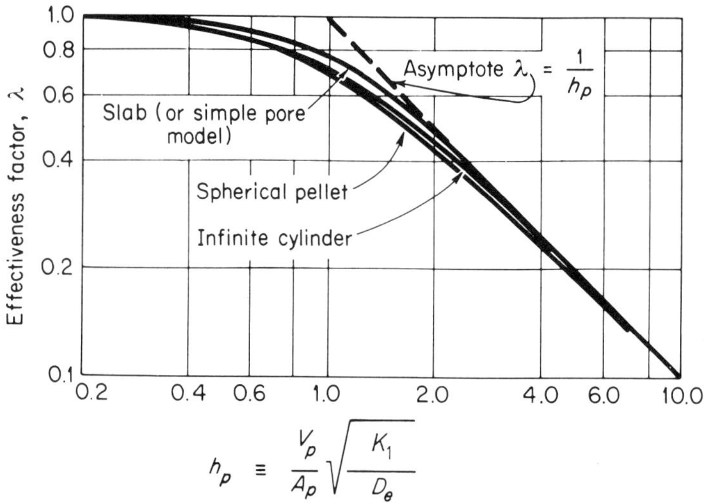

Fig. 26. Diffusion with first order reaction. Correlation of the effectiveness factors for slab, cylindrical and spherical geometrics using the generalised Thiele parameter (Petersen, 1965)

For the biological floc shown in Fig. 25 it is appropriate to define the characteristic length as:

$$L_c = \frac{V_p}{A_p} \tag{5}$$

where V_p and A_p are the volume and external surface area of a "wet" floc.

Insertion of Eq. (5) into the equations given by Atkinson and Davies (1974) leads to the biological rate equation for microbial flocs given in Table 9.

Table 9. Biological rate equation for flocs

$$R = \lambda R_{max} \frac{C^*}{K_M + C^*}$$

where, for the predominantly reaction-controlled region

$$\lambda = 1 - \frac{\tanh(k_2 V_p/A_p)}{k_2 V_p/A_p} \left(\frac{\phi}{\tanh\phi} - 1\right); \phi \leqslant 1$$

and, for the predominantly diffusion-limited region

$$\lambda = \frac{1}{\phi} - \frac{\tanh(k_2 V_p/A_p)}{k_2 V_p/A_p} \left(\frac{1}{\tanh\phi} - 1\right); \phi \geqslant 1$$

with

$$\phi = \frac{k_2 V_p/A_p}{(1 + 2C^*/K_M)^{1/2}}$$

and

$$k_2 = \left(\frac{G_{max}\rho_0}{Y_0 K_M D_e}\right)^{1/2}$$

The objectives in the development of any rate equation include:

1. the separation of the physical parameters, notably those involving concentrations and geometric factors, from those of a biochemical and microbiological nature;
2. simplicity as far as functional form is concerned;
3. a small number of kinetic coefficients;
4. universality in the sense that the same functional form of equation should be applicable to a number of substrate-microbe systems.

In the present case it is highly desirable that the rate equations for flocs and films should be algebraically similar. It is scarcely to be expected that two totally different equations would apply to the same substrate-microbe system in different geometric arrangements. This feature can be viewed as an applicability test for any proposed rate equation, whilst accepting the fact that the numerical values of the kinetic coefficients, i.e. G_{max}, K_M, and D_e, might not be exactly the same in the two situations owing to differing internal environmental conditions, e.g. pH and temperature profiles between flocs and films.

The algebraic formulations given in Table 9 are ideally suited for use with an electronic desk calculator and provide a convenient basis for the development of the algebra associated with "ideal" fermenter configurations (Atkinson, 1974).

Atkinson and Davies (1974) also provided a reduced form of the biological rate equation of Table 9. This reduced form is somewhat simpler in its algebraic formulation but is only applicable over restricted ranges of the variables.

The characteristics of the biological rate equation for flocs may be readily appreciated by reference to Fig. 27, where it is plotted as a dimensionless overall rate of substrate uptake versus dimensionless substrate concentration with the dimensionless characteristic size $(k_2 V_p/A_p)$ as a parameter. From Fig. 27 a number of asymptotes are in evidence, particular at high and low substrate concentrations. This whole family of asymptotes can be obtained readily from the equations given in Table 9.

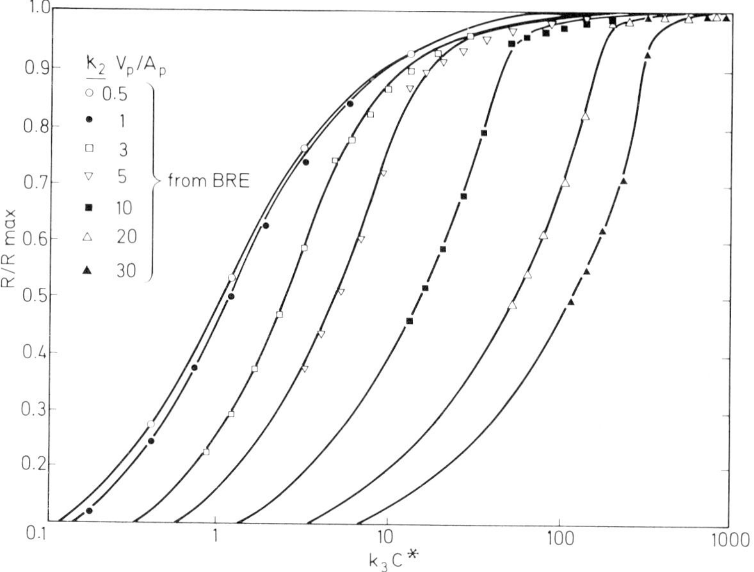

Fig. 27. Biological rate equation for flocs (Atkinson and Davies, 1974)

a) Without diffusional limitation, i.e. small V_p/A_p:

$$R = R_{\max}\frac{C^*}{K_M + C^*}\,. \tag{6}$$

b) With a potentially large diffusional limitation, i.e. large V_p/A_p:

$$R = k_2 D_e\left(\frac{A_p}{V_p\rho_0}\right)\frac{(1 + 2C^*/K_M)^{1/2}}{(1 + C^*/K_M)}C^* \text{ for small } C^* \tag{7}$$

$$R = R_{\max} \qquad \text{for large } C^*.$$

The biological rate equation for flocs exhibits most of the characteristics of potentially useful design equations; its universality is open to question because of the lack of sufficient experimental data for an extensive comparison. However, because of the generality of the underlying physical model and the known applicability of the asymptotes, there is every reason for optimism. These equations are thus intrinsically more attractive than the empirical variations of the Monod equation. The equation derived by Powell (1967) is similar in concept to the biological rate equation for flocs, but of restricted utility since it covers the "single-microorganism" situation only, and cannot be extended to multi-organism flocs. Consequently it suffers much the same limitations as the single-pore model of heterogeneous catalysis (Petersen, 1965). In contrast the biological rate equation covers the physical model which is the basis of the Powell equation (Table 8). It is advantageous to compare k_2' of the Powell model with the k_2V_p/A_p of the biological rate equation, and to note that for some organisms the effectiveness factor may always be less than unity.

Atkinson and Daoud (1968) used the experimental data of Atkinson *et al.* (1967) and Daoud (1967) to provide confirmation of the asymptotes for microbial films equivalent to those contained in Eqs. (6) and (7). For flocs the fact that the rate equation reduces to the Monod equation provides a degree of confirmation. Atkinson *et al.* (1968) have provided further confirmation of the asymptotes of the microbial film equations. Atkinson and Daoud (1970) have provided substantial evidence in support of the complete biological rate equation for films and have devised a method for the determination of the biological rate coefficients. For flocs, the work of Mueller *et al.* (1966) is particularly relevant. More recently How (1972) has used the biological rate equation to describe the experimental results of Kornegay and Andrews (1968) for microbial films and those of Yano *et al.* (1961) for flocs of *Aspergillus niger.*

5.1.1 Meaning of Parameters

Calculation of the overall rate of substrate uptake using the equations given in Table 9 requires knowledge of the biological parameters G_{max}, K_M, Y_0, and ρ_0, together with the effective diffusion coefficient (D_e). Ideally the biological parameters could be determined using flocs of single cell proportions, when the Monod equation would most probably be applicable. Caution has to be exercised in the use of Eq. (1) in case the individual organism exhibits a diffusion limitation. Knowledge of D_e is unlikely to be arrived at on the basis of simple diffusion experiments involving microbial mass in the absence of biological reaction, and consequently the BRE has to be used, together with G_{max}, K_M, Y_0, and ρ_0, to interpret data from kinetic experiments on flocs of known characteristic size. The latter requirement is difficult to achieve with any precision because of the deformable nature of the flocs, their variable water content, and not least the likelihood of a distribution of floc sizes within the apparatus. In view of these problems it seems likely that microbial films of known thickness provide the most satisfactory experimental arrangement for the determination of effective diffusion coefficients in microbial mass (Atkinson and Fowler, 1974).

The biological rate equation accounts for the lower substrate concentrations within a microbial floc which result from a diffusional limitation; it is not concerned with other factors such as differences in pH etc., which may also result from the reaction. Thus while the Monod equation may apply to the local conditions within a floc, no information is available which prescribes those conditions. To circumvent this difficulty Atkinson and Daoud (1968) introduced three biological rate equation coefficients k_1, k_2, and k_3, the values of which may depend upon the characteristic size of the microbial mass because of changes in the internal environment. In the event that the internal and external environments are identical then

$$\left.\begin{aligned} k_1 &= \frac{G_{max}\rho_0}{Y_0 K_M} \\ k_2 &= \left(\frac{k_1}{D_e}\right)^{1/2} \\ k_3 &= \frac{1}{K_M} \end{aligned}\right\} \quad (8)$$

The biological rate equation may be expressed in terms of the biological rate equation coefficients by substituting equations (8) into those given in Table 9.
The dimensionless parameter $k_2 V_p/A_p$ which occurs in the BRE and specifically in the expression for the effectiveness factor (λ) has the effect that as it is reduced, so $\lambda \rightarrow 1$.
Since λ is a measure of the diffusional limitation to substrate transfer within the floc it is clearly useful to have available a value for $k_2 V_p/A_p$ such that, for practical purposes, λ can be assumed to be unity. The data given by Atkinson and Davies (1974) suggest that when $k_2 V_p/A_p < 0.5$ then $\lambda \triangleq 1$.
Clearly for any given organism $k_2 V_p/A_p$ has a lower limit, corresponding to the single organism, which may be larger or smaller than 0.5.

5.1.2 Apparent Particle Size

The overall rate of substrate uptake by a microbial floc is related to the corresponding characteristic size by the biological rate equation of Table 9. For the equivalent problem associated with microbial films this characteristic size is clearly identifiable as the thickness of the film. Unfortunately with flocs the macromorphology is such that the volume and area of the "wet" particle are difficult to define and measure (Section 4.2). This problem is exacerbated by the fact that a particle-size distribution is likely to be a feature of any microbial suspension (Section 4.1.2). The above factors, together with those summarised in Section 4, suggest that the only useful measure of the characteristic size is likely to be one determined from kinetic considerations.
Thus the effectiveness factor of Eq. (3) can be written as:

$$\lambda = g\left(\frac{C^*}{K_M}, k_2\left(\frac{V_p}{A_p}\right)_m\right) \tag{9}$$

where $\left(\frac{V_p}{A_p}\right)_m$ represents a mean characteristic size of the flocs in the fermenter under a given set of conditions, i.e. growth rate, attrition rate, temperature, pH, composition of the medium etc.
It follows that the BRE can be used to characterise the flocs in a fermenter from a knowledge of G_{max}, K_M, Y_0, ρ_0, and D_e. Such a procedure has the advantage of being an *in situ* measurement and avoids all the problems of sampling and handling the flocs, or indeed of external measurements of questionable relevance. In fact this line of reasoning can be taken a stage further. In Section 5.1.1 knowledge of the characteristic size was referred to as a prerequisite for the evaluation of the diffusion coefficient (D_e), and the suggestion was made that this necessitated the use of a microbial film reactor. However inspection of the BRE (Table 9) suggests that since k_2 and the characteristic size are always associated as a product, it is sufficient to view the parameter $k_2 V_p/A_p$ as a dimensionless diffusional limitation and to obtain values and correlate the data accordingly.
According to this reasoning, Eq. (9) becomes:

$$\lambda = g\left(\frac{C^*}{K_M}, \left(k_2\frac{V_p}{A_p}\right)_m\right). \tag{10}$$

Use of the BRE in the form of Eq. (10) reduces the parameters to G_{max}, K_M, Y_0, ρ_0, and $(k_2 V_p/A_p)_m$. The last becomes a parameter of the microbe-substrate-fermenter system.

5.1.3 Utility of the BRE

5.1.3.1 Rate of Reaction

The biological rate equation for microbial flocs is based upon the principles of diffusion and biochemical reaction, the latter depending upon the local concentration according to the Monod relationship. The applicability of the BRE to single organisms exhibiting a diffusional limitation has already been discussed. From the above it follows that the utility of the BRE covers all organisms and limiting nutrients, i.e. carbon and nitrogen sources, oxygen, metal ions etc., providing the dependency on uptake rate follows Eq. (2), or in the event of a deviation from this equation, that this can be attributed to a diffusional limitation. It follows that the algebraic form of the BRE is of general utility and that the specific limiting nutrient-organism system is identified by the coefficients G_{max}, K_M, Y_0, ρ_0, and D_e, or k_1, k_2, k_3 whichever are appropriate.
Estimation of the rate of product formation is a more complex matter and would be expected to parallel the pattern of simple growth-associated to non-growth associated systems which characterise the single cell. In the case of simple growth-associated systems, product formation, growth, substrate and nutrient uptake would be expected to be proportional, the one to the other. For more complex systems the history of an individual organism within the floc may be important and consequently the rate of product formation may be related to floc size in a complex manner. Fig. 28 provides a speculative assessment of the relationship between rate of product formation and floc size for such a system.

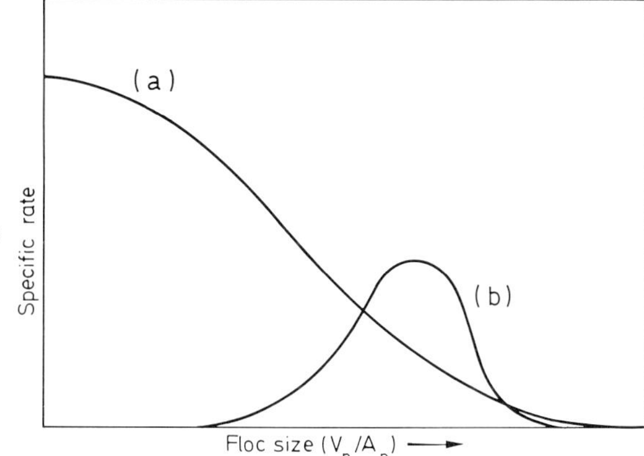

Fig. 28 a and b. Possible relationship between specific rate of product formation and floc size for a non-simply growth associated system (diagrammatic). (a) Specific rate of substrate uptake from BRE. (b) Specific rate of product formation

5.1.3.2 Pellet Growth

The increase in size of an individual pellet of spherical proportions exposed to a constant substrate concentration (C^*), neglecting any loss of mass due to attrition, is given by:

$$\frac{dm}{dt} = Gm \tag{11}$$

where m is the mass of the individual pellet.
If the substrate concentration and the particle size are such that zero order conditions prevail, i.e. $G \triangleq G_{max}$, then on integration Eq. (11) becomes:

$$m = m_0 \exp(G_{max} t) \tag{12}$$

or in terms of the biomass concentration

$$M = M_0 \exp(G_{max} t) \tag{13}$$

since

$$M = nm \tag{14}$$

where n is the number of pellets per unit volume.
Alternatively if the particle size is sufficiently large so that Eq. (7) applies, then:

$$\frac{dm}{dt} = Y_0 k_2 D_e \left(\frac{A_p}{V_p \rho_0}\right) \frac{(1 + 2C^*/K_M)^{1/2}}{(1 + C^*/K_M)} \; C^* m. \tag{15}$$

Now

$$\frac{V_p}{A_p} = \frac{d_p}{6} = \frac{1}{3} \left(\frac{3m}{4\pi\rho_0}\right)^{1/3}. \tag{16}$$

Substitution of Eq. (16) and the definition of k_2 from Table 9 into Eq. (15), followed by integration leads to:

$$m^{1/3} - m_0^{1/3} = \left(\frac{Y_0 D_e G_{max}}{K_M}\right)^{1/2} \left(\frac{4\pi}{3\rho_0^{1/2}}\right)^{1/3} \frac{(1 + 2C^*/K_M)^{1/2}}{1 + C^*/K_M} \; C^* t \tag{17}$$

or in terms of the biomass concentration

$$M^{1/3} - M_0^{1/3} = \left(\frac{Y_0 D_e G_{max}}{K_M}\right)^{1/2} \left(\frac{4\pi n}{3\rho_0^{1/2}}\right)^{1/3} \frac{(1 + 2C^*/K_M)^{1/2}}{(1 + C^*/K_M)} \; C^* t. \tag{18}$$

When $C^*/K_M \gg 1$ Eq. (18) corresponds to the so-called "cube root" law derived by Pirt (1967) Differences occur because the equation derived by Pirt is based upon the concept of diffusion with zero order reaction (Section 5.2) while Eq. (18) is based upon the region of the BRE associated with diffusion and approximately first order reaction.
A dimensionless biomass concentration can be defined as:

$$M^* = k_2 \frac{V_p}{A_p} = \frac{k_2}{3} \left(\frac{3M}{4\pi n \rho_0}\right)^{1/3} \tag{19}$$

Substitution of Eq. (19) in Eq. (18) followed by re-arrangement leads to:

$$M^* - M_0{}^* = \frac{1}{3}(G_{max}\frac{C^*}{K_M + C^*})(1 + 2C^*/K_M)^{1/2}\,t. \qquad (20)$$

The more general case involves the growth of a particle which initially is not limited by diffusion but which subsequently becomes strongly limited. For such a particle the effectiveness would decrease as a result of the increase in particle size.
In these circumstances Eq. (11) becomes:

$$\frac{dm}{dt} = Y_0\lambda R_{max}\frac{C^*}{K_M + C^*}\,m \qquad (21)$$

or

$$\int_{m_0}^{m}\frac{dm}{\lambda m} = (G_{max}\frac{C^*}{K_M + C^*})t. \qquad (22)$$

Now the effectiveness factor is given by the expression in Table 9. If the dimensionless biomass concentration of Eq. (19) is introduced into Eq. (22).

$$\int_{M_0^*}^{M^*}\frac{dM^*}{\lambda(M^*, C^*/K_M)M^*} = \frac{1}{3}\,G_{max}\frac{C^*}{K_M + C^*}\,t. \qquad (23)$$

The results of integrating Eq. (23) are given in Fig. 29 for various values of C^*/K_M. These data provide information on the time-course of the biomass concentration when the

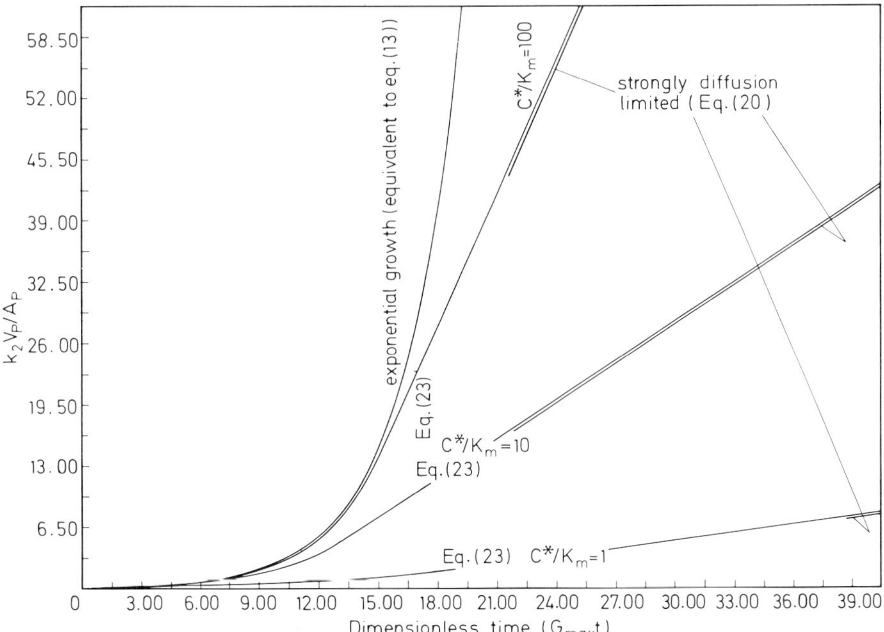

Fig. 29. Growth of particles exposed to a constant substrate concentration

particles are exposed to a constant substrate concentration. Comparisons between Eqs. (13), (20), and (23) show that the former are associated with different parts of the complete "growth" curve.

Clearly the variation with time of both the substrate concentration and biomass concentration in a batch fermenter could be simulated by using the equation

$$-\frac{dC}{dt} = \lambda RM \tag{24}$$

in conjunction with Eq. (21) (see Section 6.2.2).

5.2 Comparison between the BRE and the Zero Order Model

It is tempting, in view of the dependency on concentration indicated by Eq. (1), to approximate growth kinetics by the relationships

$$\left.\begin{array}{ll} G = G_{max} & C^* > 0 \\ G = 0 & C^* \equiv 0. \end{array}\right\} \tag{25}$$

A diagrammatic comparison between Eqs. (1) and (25) is given in Fig. 30.

The approximation is clearly valid over large ranges of concentration particularly for organism-substrate systems in which C^*_{CRIT} is low. This is a notable feature when oxygen is the rate-limiting substrate.

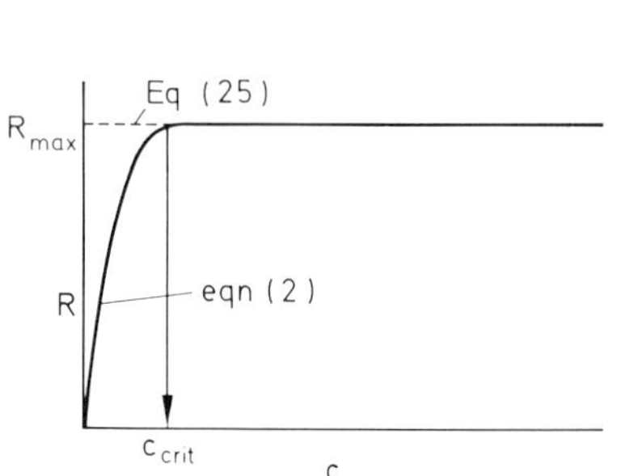

Fig. 30. Relationship between growth rate and substrate concentration

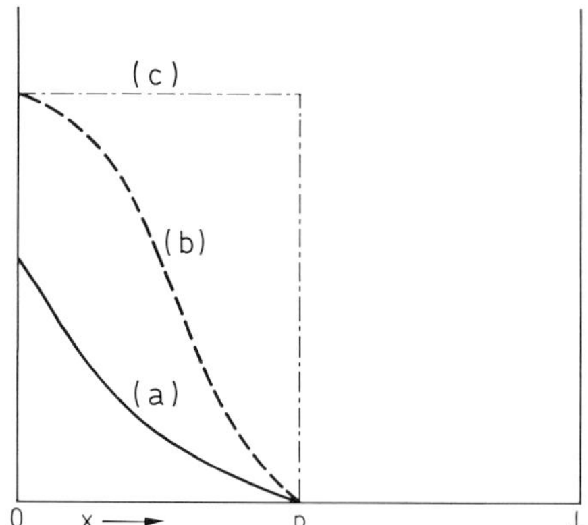

Fig. 31 a–c. Concentration and local growth rate profiles within microbial flocs. Curve (a) substrate concentration profile. Curve (b) local growth rate based upon Eq. (1). Curve (c) local growth rate based upon Eq. (25)

Unfortunately, as illustrated in Fig. 31, substantial inaccuracies can result if Eq. (25) is applied to problems involving a diffusional limitation. The overall specific growth rate of a microbial floc is given by:

$$G = \frac{1}{V_p} \int_0^{d_p/2} 4\pi G(C) r^2 \, dr. \tag{26}$$

A comparison between Eq. (26) and Fig. 31 indicates the extent of the inaccuracy involved in using Eq. (25) as compared with Eq. (1). The difference between the two integrals can be very large and this is due primarily to the manner in which substrate concentrations below C^*_{CRIT} (Fig. 30) are stretched over a large distance within the microbial mass (Fig. 31).

The above discrepancy follows because care has to be exercised to distinguish between those approximations which can only be made on the solution to the differential equation and those which may be made on the differential equation itself. In the present case only a mathematical solution to the complete problem is of any real utility.

The pragmatic attraction of using Eq. (25) lies in the fact that an analytical solution can be obtained to the problem of diffusion with "zero" order reaction (Mueller *et al.*, 1966).

This solution is given in Table 8 as

$$R = R_{\max}(1 - x_c{}^3) \tag{27}$$

where $0 < x_c \leqslant 1$.

In Eq. (27) x_c represents the depth in the floc at which the substrate concentration and therefore the growth rate falls to zero. This "critical depth" has to be calculated from the cubic equation

$$\frac{2C^*/K_M}{3(k_2 V_p/A_p)^2} = 1 - 3x_c^2 + 2x_c^3. \tag{28}$$

Equations (27) and (28) provide a further "biological rate equation" in algebraic form and a quantitative comparison can be made between this equation and the BRE to establish the extent of the inaccuracy involved in the approximation referred to above. Figure 32 contains data calculated from Eqs. (27) and (28), as well as data calculated using the BRE (Table 9). The large inaccuracies involved in using the "zero" order approximation to calculate the overall rate of substrate uptake for given values of C^*/K_M and $k_2 V_p/A_p$, can be seen clearly. The inaccuracy at low values of $k_2 V_p/A_p$ corresponds to the approximation indicated in Fig. 30 and the imposition of a diffusional limitation which does not in fact occur. At large values of $k_2 V_p/A_p$ the inaccuracy corresponds to those indicated in Fig. 31, together with an inaccurate assessment of the diffusional limitation.

5.3 Critical Substrate Concentration

The critical substrate concentration (C^*_{CRIT}) is generally described as that which leads to a near maximum growth rate, i.e. $G = n_1 G_{\max}$, where n_1 may be selected arbitrarily

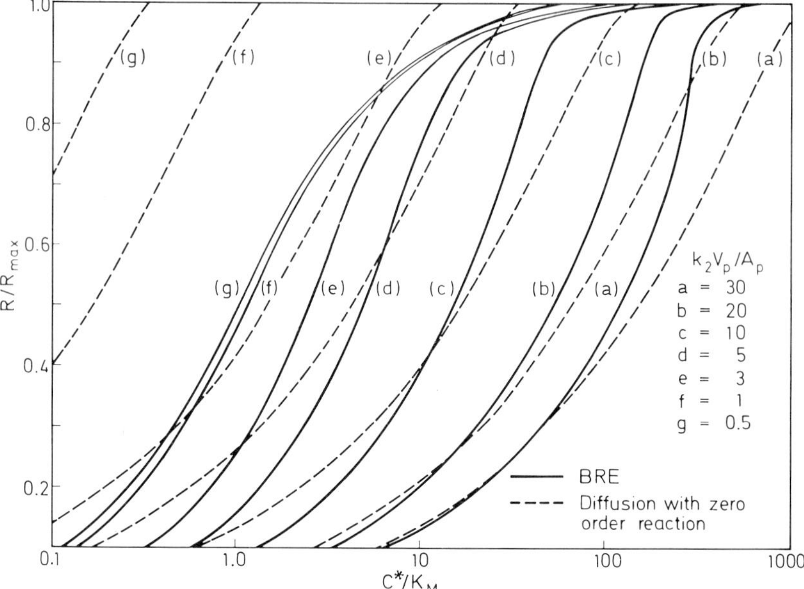

Fig. 32. Comparison between BRE (Table 9) and diffusion with zero order reaction [Eqs. (27) and (28)]

but most frequently is taken as 0.99. When the Monod equation applies, i.e. $k_2 V_p/A_p < 0.5$ (Section 5.1.1), rearrangement of Eq. (1) leads to:

$$C^*_{CRIT} = \left(\frac{n_1}{1-n_1}\right) K_M \tag{29}$$

However an appreciation of mass transfer suggests that for a diffusion-limited system C^*_{CRIT} will depend upon the particle size. Thus from the BRE (Table 9):

$$\frac{R}{R_{max}} = n_1 = \lambda\left(k_2 V_p/A_p, \frac{C^*_{CRIT}}{K_M}\right) \frac{\frac{C^*_{CRIT}}{K_M}}{1 + \frac{C^*_{CRIT}}{K_M}} . \tag{30}$$

Equation (30) can be solved in terms of C^*_{CRIT}/K_M for various values of $k_2 V_p/A_p$ and n_1. The corresponding data are given in Fig. 33, which shows the increase in C^*_{CRIT} as the particle size is increased. Equation (29) gives the limiting values of the data in Fig. 33 for small $k_2 V_p/A_p$.

When the particle size is large, $k_2 V_p/A_p > 20$ and $\phi > 1$, the effectiveness factor from Table 9 reduces to

$$\lambda_{k_2 V_p/A_p > 20} \rightarrow \frac{(1 + 2C^*/K_M)^{1/2}}{k_2 V_p/A_p} . \tag{31}$$

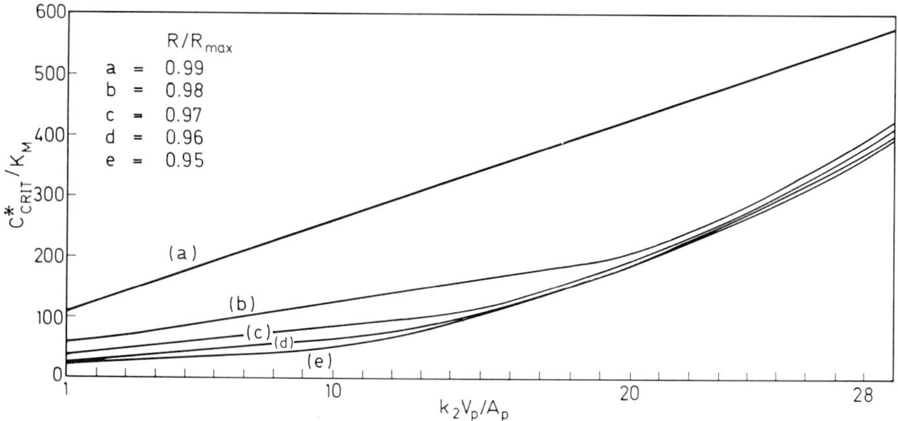

Fig. 33. Dependency of critical substrate concentration on particle size calculated from the BRE (Table 9)

Combination and rearrangement of Eqs. (30) and (31), when $(C^*/K_m) > 1$, leads to:

$$\frac{C^*_{\text{CRIT}}}{K_M} = \frac{n_1^2}{2}(k_2 V_p/A_p)^2. \tag{32}$$

Equation (32) provides an algebraic expression for large values of $k_2 V_p/A_p$.
An analytical expression for the relationship between C^*_{CRIT}/K_m and $k_2 V_p/A_p$ for various values of n_1, can also be obtained from the equations obtained for the problem of diffusion with zero order reaction (Section 5.2). From Eq. (27)

$$x_c = (1 - n_1)^{1/3} \tag{33}$$

and substitution into Eq. (28) gives

$$\frac{C^*_{\text{CRIT}}}{K_M} \simeq \frac{3}{2} n_1 (k_2 V_p/A_p)^2 \tag{34}$$

The discrepancy between Eq. (34) and the solution to Eq. (30) emphasises the dangers involved in using the zero order approximation even though it leads to algebraic equations for both the overall rate of substrate uptake and the critical substrate concentration.

5.4 Critical Particle Size

As a particle increases in size so the substrate concentration in the centre of the particle falls; concomitantly the specific rate of substrate uptake decreases. Eventually a critical particle size is reached whereby the substrate uptake is that which corresponds to an essentially zero concentration in the centre of the particle. These conditions are likely to lead to endogenous respiration and lysis in the centre of the particle and consequently it is useful to have a guide as to the magnitude of the critical particle size $(V_p/A_p)_{\mathrm{CRIT}}$.

The overall rate of substrate uptake corresponding to the critical particle size is given by

$$R_{\mathrm{CRIT}} = \lambda(k_2(V_p/A_p)_{\mathrm{CRIT}}, C^*/K_M)\,\frac{C^*/K_M}{1 + C^*/K_M} \tag{35}$$

Thus it follows that $(V_p/A_p)_{\mathrm{CRIT}}$ depends upon the substrate concentration C^*. Equation (35) can be used to evaluate $(V_p/A_p)_{\mathrm{CRIT}}$ by comparison with the BRE evaluated for $k_2 V_p/A_p > 20$, thus

$$\frac{R_{\mathrm{CRIT}}}{R(k_2 V_p/A_p \to \text{large})} = n_1 \triangleq \frac{\lambda(k_2(V_p/A_p)_{\mathrm{CRIT}}, C^*/K_M)}{\left\{\dfrac{(1 + 2C^*/K_M)^{1/2}}{k_2(V_p/A_p)_{\mathrm{CRIT}}}\right\}}. \tag{36}$$

Figure 34 contains data, calculated from equation (36), which relate $k_2(V_p/A_p)_{\mathrm{CRIT}}$ to C^*/K_M for various values of n_1. Clearly the critical size increases with increasing C^* and with the acceptable lower levels of internal concentration implicit in the larger values of n_1.

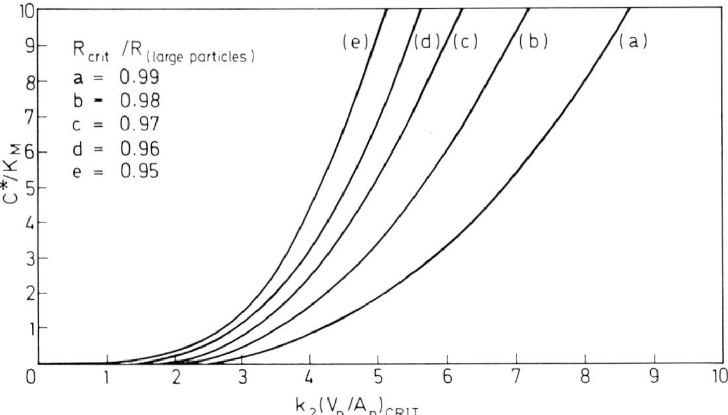

Fig. 34. Dependency of critical particle size on substrate concentration calculated from the BRE (Table 9)

6. Performance Characteristics of Fermenters Containing Microbial Flocs

6.1 Mass Transfer between Fluid and Flocs

The overall rate of substrate uptake by a microbial floc is given by:

$$R = h^1[C - C^*] = \lambda(k_2 V_p/A_p, C^*/K_m)R_{\max}\frac{C^*/K_m}{1 + C^*/K_M} \tag{37}$$

where C is the bulk of liquid concentration and h^1 is the coefficient of mass transfer (dimensions $L^3M^{-1}T^{-1}$) between the bulk liquid and the floc.
The bulk liquid concentration in Eq. (37) refers to the limiting substrate. In the case of aerobic fermentations this may be the dissolved oxygen concentration and this, in turn, is influenced by the rate of gas absorption.
Clearly to evaluate R, since C^* is unknown, it is necessary to have available information on the mass transfer coefficient h^1. Mass transfer between fluid and solid has received considerable attention in chemical engineering, e.g. Kunii and Levenspiel (1969). Unfortunately, from the biochemical engineering viewpoint, the emphasis has been on single spheres, fixed and fluidised beds, whereas the concern here is largely with 'solids' suspended either by agitators or the upflow of gas, although fluidised beds have their application as the "tower fermenter".
Equation (37) is perhaps best appreciated by reference to Fig. 35 where R is plotted diagramatically against C^* for given values of h^1, C, $k_2 V_p/A_p$, and K_m. The point of intersection of the line and curve gives the actual overall rate of reaction R and the actual interfacial concentration C^*.

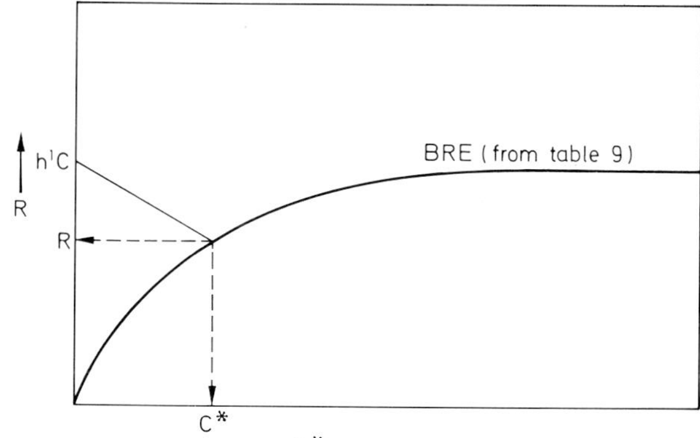

Fig. 35. Diagrammatic representation of Eq. (37) i.e. BRE in conjunction with a liquid phase diffusional limitation

6.1.1 Dispersed Suspended Solids

Data on mass transfer to single spheres immersed in a flowing fluid are correlated by the

Frössling (1938) equation

$$\frac{hd_p}{D} = 2.0 + 0.552\,Re^{1/2}Sc^{1/3} \tag{38}$$

where $Re = \frac{\rho u d}{\mu}$; $Sc = \frac{\mu}{\rho D}$

In Eq. (38) h is the mass transfer coefficient (LT^{-1}); d_p is the particle diameter (L); D is the diffusivity of the diffusing species (L^2T^{-1}); $\rho(ML^{-3})$ and $\mu(ML^{-1}T^{-1})$ are the density and viscosity of the fluid; and u is the fluid velocity (LT^{-1}).

It can be argued (Atkinson, 1971) that microbial flocs will tend to follow the fluid pattern relatively closely, since the density difference between particle and fluid is small. This factor, together with the assumption that the flocs are probably rather smaller than a turbulent "eddy", suggests that the mass transfer process might be considered as molecular diffusion into an infinite region of fluid that is stationary relative to the particle. This latter situation is described mathematically by the first two terms of Eq. (38), i.e.

$$\frac{hd_p}{D} = 2.0 \tag{39}$$

The above equation contains no dependency on the mixing device and its use leads, at the very least, to a lower limit for the mass transfer coefficient and, quite possibly, a very reasonable estimate.

The difficulty in applying Eq. (38) to dispersed suspended particles lies with the specification of the relative velocity u. On the basis of the analogous situation with "rigid" bubbles Calderbank (1967) suggested the use of the following equation

$$\frac{hd_p}{D} = 2.0 + 0.3\,Sc^{1/3}\left\{\frac{(\rho_p - \rho)d_p^3\rho g}{\mu^2}\right\}^{1/3} \tag{40}$$

where ρ_p is the particle density.

Subsequently Calderbank and Jones (1961) provided experimental support for Eq. (40) using spherical beads of ion-exchange resin in water.

The mass transfer coefficients given in Eq. (37) – (40) are related by the expression:

$$h^1 = \frac{hA_p}{V_p\rho_0}. \tag{41}$$

6.1.2 Fluidised Beds

The fluid phases passing through a bed of fluidised microbial particles may involve (a) anaerobic systems with either a single continuous liquid phase or an additional bubbling gaseous product phase (b) aerobic systems with either continuous gas and liquid phases or a continuous liquid phase with a dispersed gas phase. In any event the flow pattern of the fluid phase or phases is likely to be complex. Mass transfer in single-phase fluidised beds has received considerable attention (Satterfield, 1970) and data are given in Fig. 36.

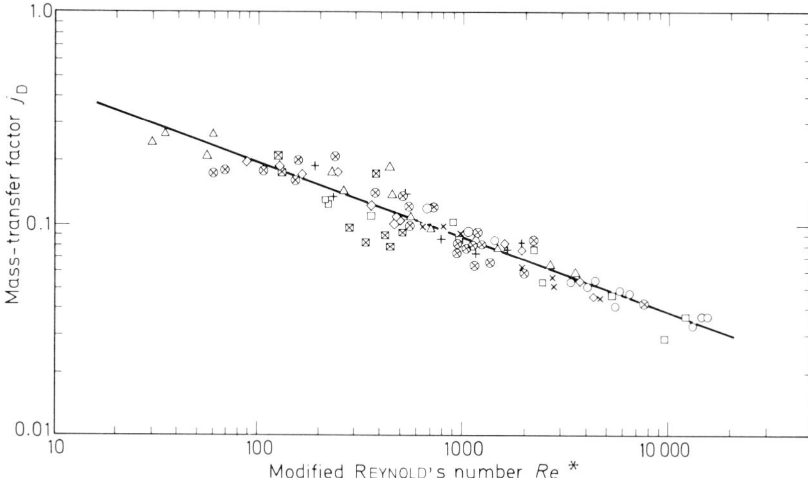

Fig. 36. Mass-transfer factor for fixed and fluidised beds–data for various gas-vapour, liquid-solute systems with spheres, and cylinders (Chu *et al.*, 1953)

$$j_D = \frac{h}{u_c} \left(\frac{\mu}{\rho D}\right)^{1/2}; Re^* = \frac{\rho u_c d_p}{\mu(1 - \epsilon_b)}$$

Unfortunately little work of a similar nature appears to have been carried out on fluid-beds involving both gas and liquid phases (Østergaard, 1968). Although little additional effect might be expected, the difficulty arises in the absence of knowledge of the hold-up of the various phases.

6.2 Batch Fermenters

Batch operation presents difficulties of interpretation of the dimensionless particle size $k_2 V_p/A_p$ exceeds 0.5 during the course of the fermentation, as this leads to a diffusional limitation (Section 5.1.1). It follows that to predict the time course of the fermentation it is necessary to have information available on the floc sizes likely to occur at various times. Unfortunately, as indicated in Sections 2, 3, and 4, the floc size is likely to be influenced by an everchanging environment. Two cases can be considered, the first when the particle size resulting from the interaction of growth, attrition and environment, remains constant even though a diffusional limitation is involved, and secondly particles which simply increase in size as a result of growth without this being counteracted either by attrition or by the changing environment.

6.2.1 Constant Particle Size

The equations describing the time course of a batch fermentation are

$$\left.\begin{aligned} \frac{dM}{dt} &= Y_0 RM \\ \text{and} \quad -\frac{dC}{dt} &= RM. \end{aligned}\right\} \tag{42}$$

If the rate of change of the conditions is not excessive then the concentration profiles in the flocs at any time may be taken to be the same as those which would exist under steady-state conditions in the same environment. Under these circumstances the substrate uptake R is given by Eq. (37).
Consideration of Eqs. (39) and (41) suggests that if $k_2 V_p/A_p < 0.5$ i.e. V_p/A_p small then the mass transfer coefficient (h^1) is likely to be large. In this case there would be neither a "solid" phase nor a liquid phase diffusional limitation and the simplifications which result lead to an analytical solution (Atkinson, 1974). If diffusional limitations exist then Eq. (42) requires numerical integration except when the liquid phase diffusion limitation is so severe that $C^* \to 0$. In this case

$$\left.\begin{aligned} \frac{dM}{dt} &= Y_0 h^1 CM \\ -\frac{dC}{dt} &= h^1 CM. \end{aligned}\right\} \tag{43}$$

Integration of Eqs. (43) yields

$$ln[\frac{C}{C_i}] - ln[\frac{\Delta - Y_0 C}{M_i}] = -h^1 t \Delta \tag{44}$$

where $\Delta = M_i + Y_0 C_i$ (45)

and M_i and C_i are the initial biomass and substrate concentrations respectively. Clearly if either a "solid" or liquid phase diffusional limitation occurs then the time required to achieve a given conversion will be increased.

6.2.2 Increasing Particle Size

The change in pellet size with time at constant concentration was discussed in Section 5.1.3.2. On the assumption that no liquid phase diffusional limitation exists, combination of Eqs. (19), (37), and (42) leads to

$$-\frac{dC}{dt} = \lambda\left\{\frac{k_2}{3}\left(\frac{3M}{4\pi n \rho_0}\right)^{1/3}, C/K_M\right\} R_{\max} \frac{C/K_M}{1 + C/K_M} M \tag{46}$$

where $M = \Delta - Y_0 C$.

Numerical integration of Eq. (46) provides the time course of both the substrate and biomass concentrations. Once more the occurence of a diffusional limitation will lead to an extended fermentation time. In contrast to Eq. (44) not only is the initial biomass concentration (M_i) required but also the number of particles.

6.3 Continuous Fermenters

6.3.1 Continuous Stirred Tank Fermenters (CSTF)

The equations describing a completely-mixed stirred tank fermenter without input of organisms or recycle are as follows

$$\left.\begin{aligned} F_M &= Y_0 RMV \\ \text{and} \quad F(C_I - C_0) &= RMV \end{aligned}\right\} \tag{47}$$

where R is given by Eq. (37).

Figure 37 provides data for a CSTF containing particles of constant size, namely $k_2 V_p/A_p$ of 30 and 0.5. The former corresponds to a strongly solid phase diffusional limitation while the latter corresponds to an absence of diffusional limitations. A considerable reduction occurs in the values of the maximum productivity and in the range of operation of the fermenter.

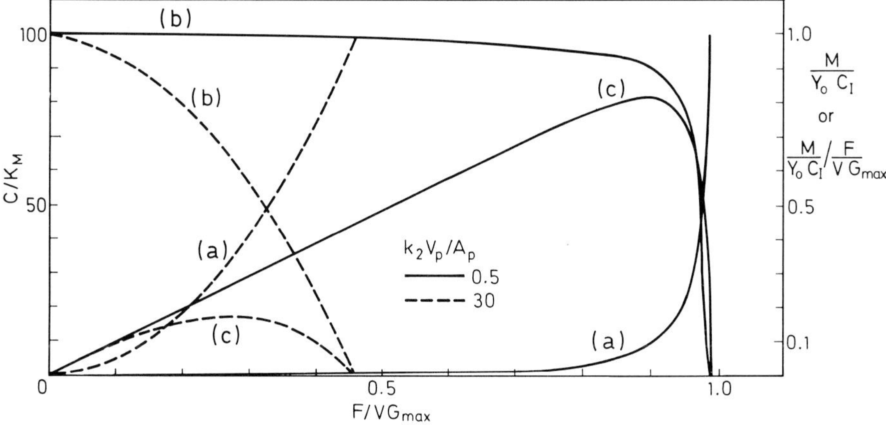

Fig. 37 a–c. Influence of floc size on the performance characteristics of a CSTF. (a) Substrate concentration. (b) Biomass concentration. (c) Productivity

The wash-out flow-rate for a CSTF is related to the particle size by evaluating Eq. (47) at C_I, i.e.

$$\left(\frac{F}{V}\right)w/o = Y_0 \lambda(k_2 V_p/A_p, C/K_M) R_{\max} \frac{C_I/K_M}{1 + C_I/K_M}. \tag{48}$$

6.3.2 Fluidised Bed Fermenters

The fluidised bed fermenter operates on the principle that if flocs of sufficient size are formed, reasonable fluid velocities can be achieved without wash-out. A crucial requirement is the attainment of a time-independent microbial hold-up and this necessitates a rate of loss of microorganisms from the fermenter equal to the growth rate. This is not easy to achieve since it depends upon a complex balance involving the particle size distribution and elutriation. The particle size distribution itself depends upon the growth rate, aggregation due to growth, and fluid mechanical attrition. In view of these factors it is hardly surprising that the most successful systems to date involve low-growth-rate brewing operations using flocculating yeasts.

To achieve a spatial concentration distribution it is obviously necessary to maintain the liquid phase in an approximation to plug-flow. There may also be advantages in very slow plug-flow as far as microbial flocs are concerned, so as to maintain them in a fixed local environment. Under these general conditions the fermenter could be described as a slowly moving packed bed. The hold-up of microbial mass in such a fermenter depends upon the liquid flow-rate and the floc sizes, as well as the density difference between microbial mass and nutrient medium. Because the density difference is quite small it is necessary to use a species and strain of microorganism that has the capacity to develop flocs of adequate size, say 1 mm or greater, to achieve finite flow-rates through the fermenter.

For a mixture of particle sizes the microbial hold-up (M) depends upon both the flow-rate and position in the fermenter (Fig. 38). The relationship developed by Richardson

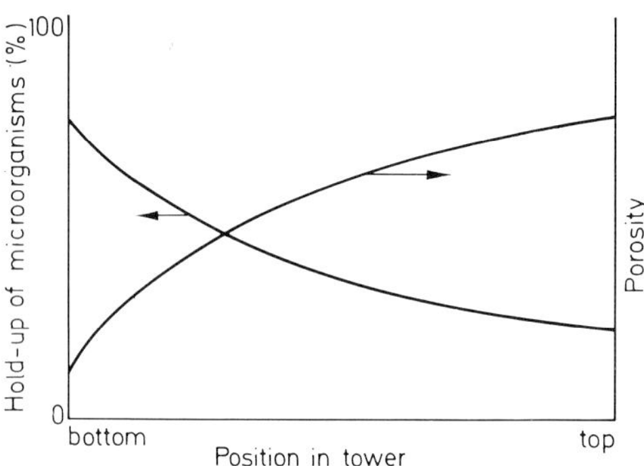

Fig. 38. Hold-up of flocs in a tower fermenter

and Zaki (1954) for a constant particle size in a fluidised bed relates the particle hold-up to the flow-rate and for the situation under consideration may be written:

$$\frac{u}{u_t} = \left(1 - \frac{M}{\rho_0}\right)^n \tag{49}$$

where n depends upon the Reynolds number ($\rho u_t d_p/\mu$), and u_t is the terminal velocity of the particles,

$$u_t = \frac{d_p^2(\rho_0 - \rho)g}{18\mu} \tag{50}$$

Since d_p and $(\rho_0 - \rho)$ are small for most fermentations the Reynolds number based on the terminal velocity will be small and near the Stokes Law region of $Re < 0.1$, where the exponent in Eq. (49) is 4.65 (Richardson and Zaki, 1954). Greenshields and Smith (1971) have shown that Eq. (49) applies to continuous beer fermentations and by using an average microbial concentration the value of n was evaluated as 4.4. This provides good agreement with the value for a constant particle size in the Stokes region of flow. Since the substrate concentrations in a fluidised bed fermenter reflect the inlet conditions quite strongly it is appropriate to assume that in many fermenters of this type zero order conditions prevail, in which case the outlet substrate concentration is given by:

$$C_0 = C_I - R_{max}\rho_0[1 - (\frac{u}{u_t})^{0.215}]\frac{Z}{Q} \tag{51}$$

In Fig. 39 are plotted the data of Klopper *et al.* (1966) in terms of the specific gravity of beer wort at the exit of a tower fermenter as a function of residence time. If the specific

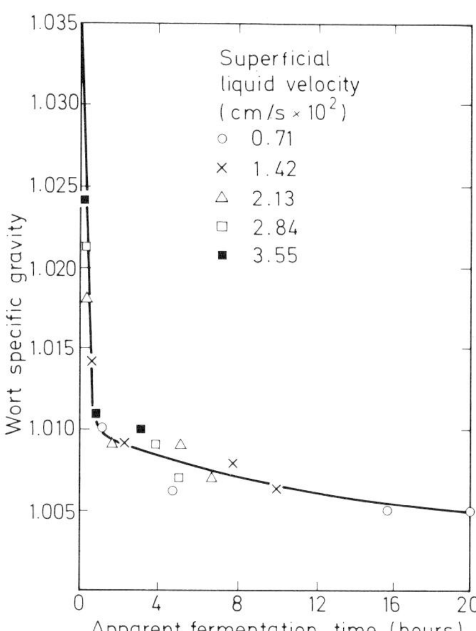

Fig. 39. Effect of liquid flow-rate on fermentation time (no allowance made for volume of yeast flocs and CO_2 bubbles) (Greenshields and Smith, 1971)

gravity is taken as a measure of substrate concentration, then to a reasonable degree of approximation the data fall on two straight lines, possibly corresponding to the fermentation of two different sugars in the wort. Thus the data are in general agreement with Eq. (51), if it is viewed as

$$SG_I - SG \propto t_f. \tag{52}$$

For beer fermentations the assumption of zero order conditions is compatible with the concentrations of fermentable sugars in grain wort, for example 90 g/l. Concentrations

of this magnitude might be expected to be in excess of the critical sugar concentration for flocs of 1 mm in size. Furthermore the microbial concentration [Eq. (49)] varies little over narrow ranges of velocity at flow-rates well below the terminal velocity.
If the particle-size distribution were available *a priori* then Eq. (49) could be applied to a series of sections in the bed. Otherwise it is necessary to assume an average value for the particle size and calculate an average hold-up.

7. Process Applications

In processes where biological matter is used the sequence of operations normally involves a biochemical reaction step followed eventually by the separation of the biological mass from the fermentation broth. In the reaction step the formation and existence of flocs may or may not be desirable, depending on the purpose and conditions of the process, and the fermenter configuration selected. Control of the environment and choice of an appropriate strain have been found to lead to the establishment of the desired form of microbial mass, i.e. floc form or otherwise. Since the presence of microorganisms as aggregates assists in the separation step, flocculation of microbial matter is often encouraged. This is carried out either in the fermentation step, when the presence of flocs is advantageous or at least not detrimental, or in a separate unit intermediate between the fermentation and separation stages.

7.1 Reaction Processes

The classic example where the presence of the floc form is advantageous for the desired fermentation to proceed is citric acid production by *A. niger*. Many workers have reported the influence of morphology on acid yield. Spherical dense pellets 0.2 – 0.5 mm in diameter have been found the most suitable to the extent of being labelled the "standard" type. The filamentous type of growth is not considered satisfactory (Steel *et al.*, 1954). Furthermore Clark *et al.* (1966) found that very soft loosely formed pellets with diameters two to three times as great as the standard type have little capacity to produce citric acid.
Another example of the use of pellets in metabolite production concerns itaconic acid; for example Nelson *et al.* (1952) used as inoculum discrete pellets 1 – 2 mm in diameter grown in shake flasks at 30° C for 2 – 3 days.
Hulme and Stranks (1971) have found that in the production of cellulase by *Myrothecium verrucaria* the organism developed as small pellets and the nutritional status of each cell depended on its position in the pellet. They report that in cultures with larger pellets, the individual cells were in different stages of development and results were generally erratic.
The above represents the positive benefits on the extent of product formation which can result from the pellet morphology. Clearly such an effect is pre-eminent but against this has to be set the reduction in the overall rate of substrate uptake which results from the

reduced substrate concentration within the floc due to the diffusional limitation (Section 5). In the event that the flocs are produced to aid biomass separation this advantage has to be set against the reduced uptake rate.

7.2 Physical Processes

The form of the microbial mass in fermentation has a direct effect both on the physical properties of the broth and on the physical processes taking place, i.e. on the viscosity, on aeration, fouling and the subsequent separation of the biomass.
In the case of fungal microorganisms a considerable difference is found in the rheological properties of fermentation broth between mycelial and pellet growth. Pirt and Callow (1959) have suggested that from the practical point of view the long hyphae of *Penecillium chrysogenum* produced at low pH are undesirable because they make the viscosity of the culture high and consequently lower the efficiency of mixing and the rate of transfer of oxygen. The effect of the morphology of *A. niger* on the rheology of the broth has been studied by Morris *et al.* (1973). The oxygen transfer rate was decreased substantially by the use of high densities of filamentous mould mycelium. This was reflected in a very high viscosity of the fermentation broth. Furthermore changes from filamentous pellets to smooth rounded pellets also resulted in changes in viscosity (Fig. 40). These changes were further reflected in the microbial hold-up and in diffe-

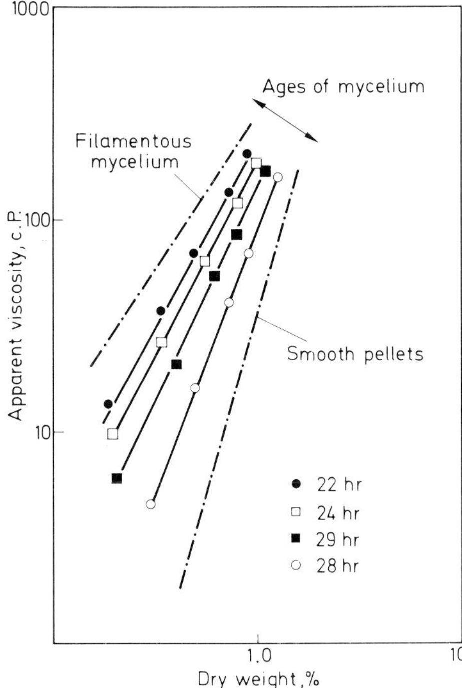

Fig. 40. Changes in the viscosity of *A. niger* pellets during growth in a tower fermenter (Morris *et al.*, 1973)

rences in the gas hold-up (Fig. 41). Clark and Lentz (1963) have fermented beet molasses with mechanically crushed pellets of *A. niger* in a tower fermenter and found that the viscosity of the mash after 90 hrs of fermentation was usually greater than 400 centipoise.

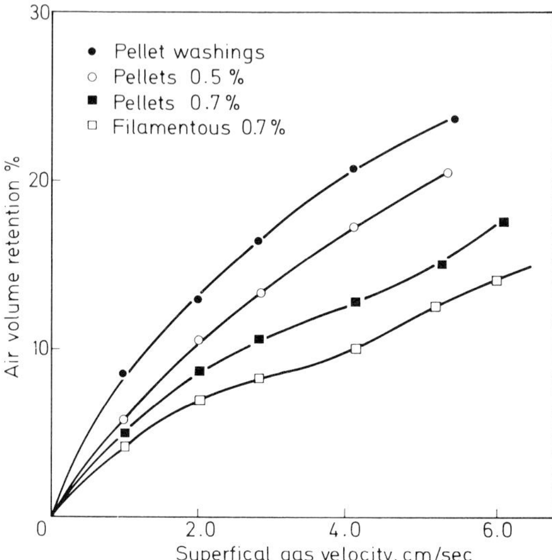

Fig. 41. Air hold-up for pelleted and filamentous mycelium of *A. niger* in a tower fermenter (Morris *et al.*, 1973)

It has also been reported that the use of fungal pellets may eliminate the slime-forming tendencies of the fungus. Litchfield (1967), when growing mushroom mycelium (*Morchella hortensis*) in submerged culture, found that the form of growth depended on the aeration-agitation system used, and that filamentous mycelial growth tended to wrap around the impeller and to foul the agitator blades.

The removal of cellular material from fermentation broths is important in fermentation processes. This is so, regardless of the objectives of the process. In different processes microbial flocs may be found to be present throughout the fermentation stage, form at the end of it as a matter of course, or be encouraged to form through a change in the fermenter environmental conditions, e.g. by the use of chemical flocculants. Examples where aggregates are used in stirred tank fermenters include: citric acid, penicillin and mushroom production. Aggregates are generally used in tubular forms of fermenter, and such systems have found application for beer, alcohol, malt vinegar, citric acid and biomass production (Greenshields and Smith, 1971; Smith and Greenshields, 1974). Clearly biological aggregation provides a convenient and effective method for separation of microbial species from their suspending medium after they have fulfilled their metabolic role. Sections 2, 3, and 4 suggest that the environmental conditions should be adjusted, at the end of the product formation phase, so as to be favourable to aggregation. In the activated sludge process, flocculation is desirable after the fermentation step is completed. Indeed it has been suggested by Tenny and Stumm (1965) and Parker *et al.*

(1971) that a flocculation stage be included between the aeration and the secondary clarification stages, i.e. following the biosynthesis step. The activated sludge process inherently relies on two independent characteristics for the production of an acceptable effluent. The first is the complete assimilation of dissolved, suspended, and colloidal organic material by the active mass of microorganisms to final end-products of carbon dioxide, water, inert material and biomass. This involves substrate utilization and active synthesis of microbial mass. The second phase is the flocculation of the microorganisms and other suspended or colloidal components into readily settleable flocs, so that a clear, low-BOD end-product may be obtained. The highly turbulent conditions in the aerator lead to floc break-up and it has been suggested (Parker *et al.* 1971) that an environment more conducive to aggregation and the incorporation of dispersed particles into flocs, could be obtained using mild stirring.

With respect to brewing, Rainbow (1966) states that, while the biochemical activity of yeast in producing ethanol, carbon dioxide and the minor products which contribute to the character and flavour of beer, is the most important aspect in the selection of yeast, the property of flocculence is also important. This property determines the formation of flocs which sediment in bottom fermentations (lager) and float in top fermentations (ale). In the primary stage of bottom fermentation some breweries use a mixture of sedimentary (flocculent) and powdery (non-flocculent) strains of yeast, while other breweries select a strain which, after a normal period of sedimentation, leaves the correct content of yeast and fermentable matter in the beer for secondary fermentation. Failing this, a flocculent yeast is used for fermenting a proportion of wort while the remainder is pitched with a non-flocculent strain; the two beers are then mixed in the lagering tanks, to ensure adequate yeast and fermentable material, (Hough *et al.* 1971). The traditional lagering process (secondary fermentation) extends over several months during which the remaining yeast settles out.

In the course of a typical bottom fermentation the temperature rises gradually from 6° C to 10° C in the first 6 days, then it falls gradually to reach 5° C on the 10th day. The last two-day period is referred to as the cooling and settling stage. A reduction in temperature 5° C to 0° C is used to help settlement in the lagering tanks.

Pirt and Callow (1959) proposed a two-stage continuous flow process for penicillin production. The optimum pH for penicillin production is about 7.4 and at values greater than 7.0 there is extensive formation of pellets and the swollen aberrant form. A first stage is required for growth of the mould with the pH value not exceeding 7.0 and a second stage with a higher pH value for penicillin production.

Chemical flocculation has been suggested for the activated sludge process and for biomass production; it is already in use in beer clarification. Tenney and Stumm (1965) advocated the use of chemical flocculation in the activated sludge process suggesting that this would allow the design of the biological reactor to be optimised without regard to the separation stage.

For biomass production Gasner and Wang (1970) have suggested microbial cell recovery enhancement through flocculation. Johnson (1967) has suggested that if the centrifugal harvesting of bacteria could be improved by flocculation, then bacteria would have an important advantage over yeast in protein production because of their versatility in substrate utilization. Cationic polyamine and positively charged alumina fibrils have

been tested by McGregor and Finn (1969) as flocculants for the bacteria, *Pseudomonas fluorescens, Escherichia coli* and *Lactobacillus delbruckii.*
In beer clarification, finings (solubilized collagen and gelatin) are used; the molecules have both positive and negative charges but the overall charge is positive at the pH of beer. These agents react very effectively with the yeast cells, whose overall charge is negative, and lead to the aggregation of yeast cells and finings, and the larger particles so formed sink more readily (Hough *et al.*, 1971).

7.3 Fermenter Configuration and Microbial Hold-up

The conventional fermenter configuration is the stirred tank. Under continuous conditions such fermenters have a limited range of operation owing to wash-out of the biomass and the microbial hold-up is largely dictated by the throughput. Attempts to increase the microbial hold-up and therefore productivity per unit volume, to increase the range of operation, and to produce a situation in which the biomass is exposed to a variety of environments, have led to a number of developments. In continuous fermentation the use of biomass recycle following a sedimentation stage leads to both increased microbial hold-up and an extended range of operation (Atkinson, 1974); an internal sedimentation stage has the same effect (Pirt and Kurowski, 1970). Martin and Waters (1952) have employed a tower type fermenter in their studies of the production of citric acid by *A. niger* pellets, whilst Greenshields and Smith (1971) have reported studies on anaerobic and aerobic fermentations in tower fermenters. Atkinson (1974) proposed a fluidised-bed fermenter and presented a mathematical description of the probable performance characteristics. All these configurations are aided by the presence of flocs since, by and large, this leads to increased microbial hold-up for a given set of conditions.
Under batch conditions the presence of flocs may lead to enhanced product formation and a reduced viscosity. However, James (1973) found that the total liquid content of *A. niger* ranged from 79% by volume for mycelia to 98.5% for indivudal pellets containing immobile liquid.
The dry matter per unit volume of suspension equals

$$\rho_s(1-\epsilon_b)(1-A) \tag{53}$$

where ϵ_b is the liquid fraction excluding flocs;
A is the liquid fraction within the flocs and
ρ_s is the density of the dry matter.

Using the above equation, with the assumption that the minimum total liquid content to be found in pellets will be 79%, the maximum dry mycelial weight fraction possible was calculated to be 4.7 gm%. James (1973) reported an experimental maximum yield of 2 gm%. Because of the increased water content with the larger particles the microbial hold-up is greatest when single cells are present. Equation (53) can be used to compare biomass hold-up, under batch conditions, for various values of suspension voidage and total liquid content (Fig. 42). In the case of continuous flow in a tower fermenter the effect of particle diameter on biomass hold-up at various flow rates is given in Fig. 43.

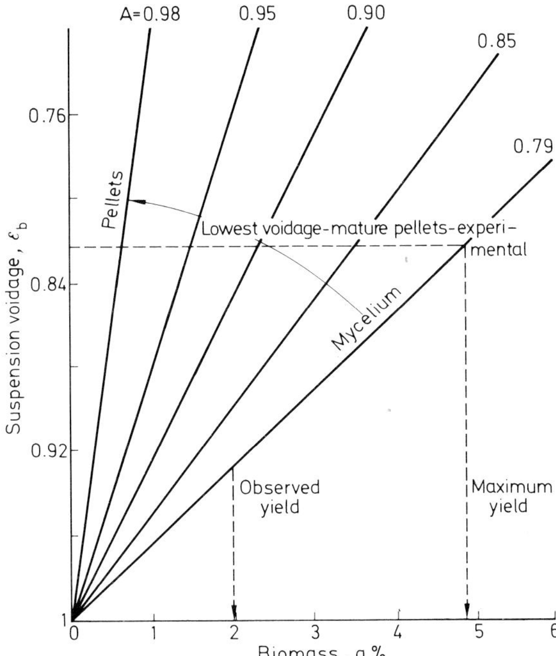

Fig. 42. Batch biomass hold-up for *A. niger* (James, 1973) (A = liquid fraction within microbial pellets)

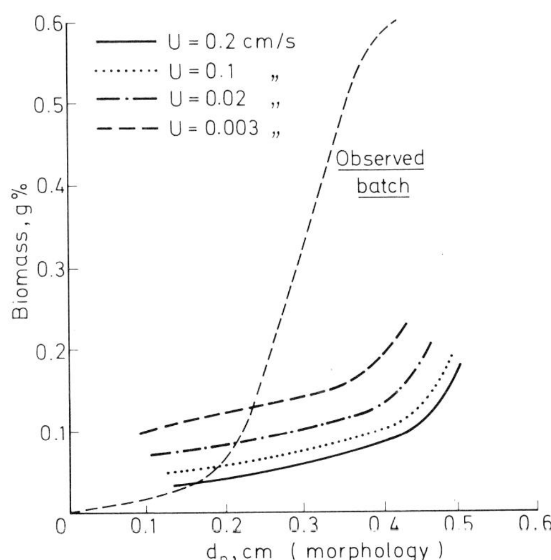

Fig. 43. Experimental data on biomass hold-up for continuous *A. niger* production in a tower fermenter (James, 1973)

Nomenclature

A	=	liquid fraction within microbial flocs	–
A_p	=	area of a single "wet" biological particle or floc	L^2
B_i	=	empirical coefficient used in Table 8	various
C	=	limiting substrate concentration	ML^{-3}
C^*	=	limiting-substrate concentration at interface between 'solid' microbial mass and the adjacent solution	ML^{-3}
C_i	=	initial concentration of limiting substrate	ML^{-3}
C_I	=	inlet substrate concentration	ML^{-3}
C_0	=	outlet substrate concentration	ML^{-3}
C^*_{CRIT}	=	critical substrate concentration (defined in Fig. 30)	ML^{-3}
d_p	=	particle diameter	L
D	=	liquid-phase diffusion coefficient	L^2T^{-1}
D_e	=	effective diffusion coefficient within microbial mass	L^2T^{-1}
F	=	flow-rate	L^3T^{-1}
$F_{w/o}$	=	flow-rate at wash-out	L^3T^{-1}
g	=	gravitational constant	LT^{-2}
G	=	specific growth-rate	T^{-1}
G_{max}	=	maximum specific growth-rate	T^{-1}
h	=	liquid-phase mass transfer coefficient [defined by Eq. (38)]	LT^{-1}
h^1	=	liquid-phase mass transfer coefficient [defined by Eq. (37)]	$L^3M^{-1}T^{-1}$
k_1	=	biological rate equation coefficient [defined by Eq. (8)]	T^{-1}
k_2	=	biological rate equation coefficient (defined in Table 9)	L^{-1}
k_3	=	biological rate equation coefficient [defined by Eq. (8)]	$M^{-1}L^3$
k'_2	=	dimensionless biological rate equation coefficient used in Table 8	–
K_M	=	Monod coefficient	ML^{-3}
L_c	=	characteristic length	L
m	=	mass of an individual microbial particle	M
m_0	=	initial mass of an individual microbial particle	M
m'	=	dimensionless parameter (defined in Table 8)	–
M	=	microbial mass concentration	ML^{-3}
M_i	=	initial microbial mass concentration	ML^{-3}
M_0	=	initial microbial mass concentration	ML^{-3}
M^*	=	dimensionless biomass concentration [defined by Eq. (19)]	–
M^*_0	=	initial dimensionless biomass concentration	–
n	=	number of microbial particles per unit volume	L^{-3}
n_1	=	a number near but less than unity	–
N	=	flux of substrate at the interface between microbial mass and adjacent solution	$ML^{-2}T^{-1}$
N_{max}	=	maximum flux of substrate at interface between microbial mass and adjacent solution	$ML^{-2}T^{-1}$
Q	=	volumetric throughput per unit area	LT^{-1}
R	=	specific rate of substrate removal by a microbial floc (rate of removal per unit microbial mass)	T^{-1}

R_{CRIT}	=	overall rate of substrate uptake corresponding to zero substrate concentration in centre of particle	T^{-1}
R_{max}	=	maximum specific rate of substrate removal by a microbial floc (rate of removal per unit microbial mass)	T^{-1}
SG	=	specific gravity of wort	–
SG_I	=	inlet specific gravity of wort	–
t	=	time	T
u	=	fluid velocity	LT^{-1}
u_c	=	superficial or open-tube velocity	LT^{-1}
u_t	=	terminal velocity of a particle [defined by Eq. (50)]	LT^{-1}
V	=	fermenter volume	L^3
V_p	=	volume of a single "wet" biological particle or floc	L^3
$(V_p/A_p)_{CRIT}$	=	critical particle size [defined by Eq. (35)]	L
x_c	=	dimensionless penetration depth (defined in Table 8)	–
Y_0	=	yield coefficient	–
Z	=	fermenter length	L
Δ	=	equivalent inlet concentration of microbial mass [defined by Eq. (45)]	ML^{-3}
ϵ_b	=	porosity of bed	–
λ	=	effectiveness factor [defined by Eq. (3)]	–
λ_i	=	effectiveness factor (defined in Table 8)	–
μ	=	liquid viscosity	$ML^{-1}T^{-1}$
ρ	=	liquid density	ML^{-3}
ρ_0	=	density of cells measured in terms of *dry* weight per unit *wet* volume	ML^{-3}
ρ_p	=	particle density	ML^{-3}
ρ_s	=	density of dry matter	ML^{-3}
ϕ	=	dimensionless parameter (defined in Table 9)	–
ψ	=	dimensionless parameter (defined in Table 8)	–

References

Aiba, S., Nagatani, M.: Advan. Biochem. Eng. **1**, 31 (1971).

Atkinson, B.: Biochemical Reaction Engineering in Chemical Engineering, Vol. III. Coulson, J. M.: Richardson, J. F. (Eds.). Pergamon Press, Oxford. 347 (1971).

Atkinson, B.: Biochemical Reactors. London: Pion Press (1974).

Atkinson, B., Busch, A. W., Swilley, E. L., Williams, D. A.: Trans. Inst. Chem. Engrs. (London) **45**, T 257 (1967).

Atkinson, B., Daoud, I. S.: Trans. Inst. Chem. Engrs. (London) **46**, T 19 (1968).

Atkinson, B., Daoud, I. S.: Trans. Inst. Chem. Engrs. (London) **48**, T 245 (1970).

Atkinson, B., Daoud, I. S., Williams, D. A.: Trans. Inst. Chem. Engrs. (London) **46**, 245 (1968).

Atkinson, B., Davies, I. J.: Trans. Inst. Chem. Engrs. (London) **52**, 248 (1974).

Atkinson, B., Fowler, H. W.: Advan. Biochem. Eng. **3**, 221 (1974).

Baig, M. A., *et al.*: Pakistan J. Sci. Ind. Res. **15**, 58 (1972). Quoted by Whitaker and Long (1973).

Baker, D. A., Kirsop, B. H.: J. Inst. Brew. 78, 454 (1972).

Barber, R. T.: Nature. **211**, 257 (1966).

Beetlestone, H. C. J.: J. Inst. Brew. **36**, 483 (1930).

Blakebrough, N., Hamer, G.: Biotechnol. Bioeng. **5**, 59 (1963).

Block, S. S., Stearns, T. W., Stephens, R. L., McCandless, R. F. J.: Agr. Food Chem. **1**, 890 (1953).

Bower, J. E., Bates, R. G.: J. Res. Nat. Bur. Stand. **55**, 197 (1955). Quoted by Stanley and Rose (1967).

Calderbank, P. H.: In Biochemical and Biological Engineering Science, Vol. I. Blakebrough, N. (Ed.). New York: Academic Press 1967.

Calderbank, P. H., Jones, S.: Trans. Inst. Chem. Engrs. **39**, 363 (1961).

Calleja, G. B.: J. Gen. Microbiol. **64**, 247 (1970).

Choudhary, Q., Pirt, S. J.: J. Gen. Microbiol. **41**, 99 (1965).

Chu, J. C., Kahil, J., Wetteroth, W. A.: Chem. Eng. Prog. **49**, 141 (1953).

Clark, D. S.: Biotechnol. Bioeng. **4**, 17 (1962a).

Clark, D. S.: Cand. J.: Microbiol. **8**, 133 (1962b).

Clark, D. S., Ito, K., Horistu, H.: Biotechnol. Bioeng. **8**, 465 (1966).

Clark, D. S., Lentz, C. P.: Biotechnol. Bioeng. **5**, 193 (1963).

Clark, J. B.: J. Bacteriol. **75**, 400 (1958).

Cocker, R., Greenshields, R. N.: Inst. Chem. Engrs. Ann. Res. Meeting. Biochem. Eng. Sec. University of Bradford. March 1975.

Cohen, G. N., Monod, J.: Bacteriol. Rev. **21**, 169 (1957).

Contois, D. E.: J. Gen. Microbiol. **21**, 40 (1959).

Conway, E. J., Downey, M.: Biochem. J. **47**, 347 (1950).

Coulson, J. M., Richardson, J. F.: Chemical Engineering Vol. I. 2nd Edn. Pergamon Press. Oxford (1964).

Coulson, J. M., Richardson, J. F.: Chemical Engineering Vol. II. 2nd. Edn. Pergamon Press. Oxford (1968).

Daoud, I. S.: M. Sc. Thesis. University of Wales (1967).

DeWalle, F. B., Chain, E. S. K.: Biotechnol. Bioeng. **16**, 739 (1974).

Dixon, L. K., Lamer, V. K., Cassian, L. I., Messinger, S., Linford, H. B.: J. Colloid Sci. **23**, 465 (1967).

Eddy, A. A.: J. Inst. Brew. **61**, 313 (1955).

Eddy, A. A.: J. Inst. Brew. **64**, 143 (1958).

Eddy, A. A., Rudin, A. D.: J. Inst. Brew. **64**, 368 (1958).

Foster, W. J.: Chemical Activities of Fungi. Academic Press (1949).

Frössling, N.: Gerland Beitr. Geophys. **52**, 170 (1938).

Gasner, L. L., Wang, D. I. C.: Biotechnol. Bioeng. **12**, 127 (1970).

Gierer, A.: Sci. Amer. **231**, 44 (1974).

Gomori, G.: Meth. Enzynol. **1**, 138 (1955). Quoted by Stanley and Rose (1967).

Graebe, J. E., Novelli, G. D.: Exp. Cell. Res. **41**, 509 (1966).

Green, M. B., Cooper, B. E., Jenkins, S. H.: Air Water Pollution **9**, 807 (1965).

Greenshields, R. N., Smith, E. L.: Chem. Engr. No. **249**, 182 (1971).

Harris, R. H., Mitchel, R.: Ann. Rev. Microbiol. **27**, 27 (1973).

Hawker, L. E., Linton, A. H., Flokes, B. F., Carlile, M. J.: London Arnold (1963).

Healy, T. W.: J. Colloid Sci. **16**, 609 (1961).

Hodge, H. M., Metcalfe, S. N.: J. Bacteriol. **75**, 258 (1958).

Hough, J. S.: J. Inst. Brew. **63**, 483 (1957).

Hough, J. S., Briggs, D. E., Stevens, R.: Malting and Brewing Science. Chapman and Hall Ltd. (1971).

How, S. Y., Ph. D. Thesis. University of Wales (1972).

Hulme, M. A., Stranks, D. W.: J. Gen. Microbiol. **69**, 145, (1971).

James, A.: Ph. D. Thesis. University of Aston (1973).

James, A., Smith, E. L.: Inst. Chem. Engrs. Ann. Res. Meeting. Biochem. Eng. Sec. University of Bradford. March (1975).

Johnson, M. J.: Science. **155**, 1515 (1967).

Klopper, W. J., Roberts, R. H., Royston, M. G., Ault, R. G.: Proc. 10th Congress Eur. Brew Conv. Elsevier. Amsterdam (1966).

Kobayashi, T., Van Dedem, G., Moo-Young, M.: Biotechnol. Bioeng. **15**, 27 (1973).

Kornegay, B. H., Andrews, J. F.: J. Water Pollution Control Federation. **40**, 460 (1968).

Kunii, D., Levenspiel, O.: Fluidization Engineering. John Wiley (1969).
Lindquist, W. J.: J. Inst. Brew. **59**, 59 (1953).
Linke, W. F., Booth, R. B.: Trans. Am. Inst. Mining (Metallurgical) Engrs. **217**, 364 (1960).
Litchfield, J. H.: 'Submerged Culture of Mushroom Mycelium' in Microbial Technology. Peppler, H. J. (Ed.). Reinhold 107 (1967).
Litchfield, J. H., Overbeck, R. C., Davidson, R. S.: Agr. Food, Chem. **11**, 158 (1963).
Lyons, T. P., Hough, J. S.: J. Inst. Brew. **76**, 564 (1970).
Lyons, T. P., Hough, J. S.: J. Inst. Brew. **77**, 300 (1971).
Mallette, M. F.: Evaluation of Growth by Physical and Chemical Means. In: Methods in Microbiology. Vol. I. Norris, J. R., Ribbons, D. W. (Eds.). New York. Academic Press **521** (1969).
Manary, O. J.: Quoted by Wallas (1968) as private communication.
Mandels, M.: The Culture of Plant Cells. In: Adv. Biochem. Eng. **2**, 202 (1972).
Martin, S. M., Waters, W. R.: Ind. Eng. Chem. **44**, 2229 (1952).
Masschelein, C. A., Devreux, A.: Proc. Eur. Brew. Conv. Elsevier 194, (1957).
McGregor, W. C., Finn, R. K.: Biotechnol. Bioeng. **11**, 127 (1969).
McIlvaine, T. C.: J. Biol. Chem. **49**, 183 (1921). Quoted by Stanley and Rose (1967).
McKinney, R. E.: In: 'Biological Treatment of Sewage and Industrial Wastes'. Vol. I. McCabe, J., Eckenfelder, W. W. (Eds.). Reinhold (1956).
McWilliam, A. A., Smith, S. M., Street, H. E.: Ann. Bot. **38**, 243 (1974).
Michaels, A. S.: Ind. Eng. Chem. **46**, 1485 (1954).
Mill, P. J.: J. Gen. Microbiol. **35**, 53 (1964a).
Mill, P. J.: J. Gen. Microbiol. **35**, 61 (1964b).
Mill, P. J.: J. Gen. Microbiol. **44**, 329 (1966).
Millis, N. F., Trumpy, B. H., Palmer, B. M.: J. Gen. Microbiol. **30**, 365 (1963).
Monod, J.: Ann. Rev. Microbiol. **3**, 371 (1949).
Morris, G. G., Greenshields, R. N., Smith, E. L.: Biotechnol. Bioeng. Symp. No. **4**, 535 (1973).
Moscona, A. A.: Sci. Amer. **205**, 142 (1961a).
Moscona, A. A.: Nature. **190**, 408 (1961b).
Moscona, A. A., Moscona, M. H.: Exp. Cell Res. **41**, 642 (1966).
Moser, H.: The Dynamics of Bacterial Populations in the Chemostat. Carnegie Inst. Publication No. 614. Washington D. C. (1958).
Mueller, J. A., Voelkel, K. G., Boyle, W. C.: J. Sanit. Eng. Div. Am. Soc. Civil Engrs. SA2. **92**, 9 (1966).
Nelson, G. E. N., Traufler, D. H., Kelley, S. E., Lockwood, C. B.: Ind. Eng. Chem. **44**, 1166 (1952).
Novick, A.: Ann. Rev. Microbiol. **9**, 97 (1955).
Østergaard, K.: Advan. Chem. Eng. **7**, 71 (1968).
Parker, D. S., Kaufman, W. J., Jenkins, D.: J. Water Poll. Control Fed. **43**, 1817 (1971).
Passmore, S. M.: J. Inst. Brew. **79**, 237 (1973).
Pavoni, J. L., Tenney, M. W., Echelberger, W. F.: J. Water Poll. Control. Fed. **44**, 414 (1972).
Perret, C. J.: J. Gen. Microbiol. **22**, 589 (1960).
Petersen, E. E.: Chemical Reaction Analysis. Prentice-Hall. Englewood Cliffs. New Jersey (1965).
Pirt, S. J.: Proc. Roy. Soc. London. Series B. **166**, 369 (1967).
Pirt, S. J., Callow, D. S.: Nature. **184**, 307 (1959).
Pirt, S. J., Kurowski, W. M.: J. Gen. Microbiol. **29**, 233 (1970).
Powell, E. O.: Growth Rate of Microorganisms as a Function of Substrate Consumption. In: Microbial Physiology and Continuous Culture. Powell, E. O., Evans, C. G. T., Strange, R. E., Tempest, D. W. (Eds.). 3rd Int. Symp. HMSO, London 34 (1967).
Puhan, Z., Martin, S. M.: Prog. Ind. Microbiol. **9**, 13 (1971).
Rainbow, C.: Proc. Biochem. **1**, 489 (1966).
Renn, C. F.: In: Biological Treatment of Sewage and Industrial Wastes. Vol. I. McCabe, J., Eckenfelder, W. W. (Eds.), Reinhold (1956).
Richardson, J. F., Zaki, W. N.: Trans. Inst. Chem. Engrs. **32**, 35 (1954).
Satterfield, C. N.: Mass Transfer in Heterogeneous Catalysis. MIT Press. Cambridge, Mass. (1970).
Smith, E. L., Greenshields, R. N.: Chem. Engr. No. 281, 28 (1974).

Smith, S. M., Street, H. E.: Ann. Bot. **38**, 223 (1974).
Smith, W. E., Volk, H.: Quoted by Wallas (1968) as private communication.
Stanley, S. O., Rose, A. H.: J. Gen. Microbiol. **48**, 9 (1967).
Steel, R., Lentz, C. P., Martin, S. M.: Can. J. Microbiol. **1**, 299 (1955).
Steel, R., Martin, S. M., Lentz, C. P.: Can. J. Microbiol. **1**, 150 (1954).
Taylor, N. W., Orton, W. L.: J. Inst. Brew. **79**, 294 (1973).
Tenney, M. W., Stumm, W.: J. Water Pollut. Contr. Fed. **37**, 1370 (1965).
Tessier, G.: Rev. Sci. **80**, 209 (1942). Paris.
Thomas, J. M., Thomas, W. J.: Introduction to the Principles of Heterogeneous Catalysis. Academic Press. New York (1967).
Thorne, R. S. W.: Proc. Euro. Brew. Conv. 21 (1951).
Trinci, A. P. J.: Arch. Mikrobil. **73**, 353 (1970). Quoted by Whitaker and Long (1973).
Walles, W. E.: J. Colloid Sci. **27**, 797 (1968).
Wang, D. I. C., Humphrey, A. E.: Chemical Eng. **76**, 108 (1969).
Whitaker, A., Long, P. A.: Process. Biochem. **8**, No. 11, 27 (1973).
Yano, T., Kodama, T., Yamada, K.: Agr. Biol. Chem. **25**, 580 (1961).
ZoBell, C. E.: J. Bacteriol. **46**, 39 (1943).

CHAPTER 3

Analog/Hybrid Computation in Biochemical Engineering

P. L. ROGERS,
The University of New South Wales, P.O. Box 1, Kensington, New South Wales, Australia 2033

With 18 Figures

Contents

1. Introduction

Recent reviews have discussed the role of computers in biochemical engineering, and their application to simulation and mathematical modelling (Nyiri, 1972; Blanch and Dunn, 1974). Both off-line and on-line digital computation techniques have been evaluated and some mention has been made also of the available analog computational methods. However, the latter subject has not been fully explored nor any attempt made to assess limitations and possible advantages of the analog/hybrid approach. Furthermore, as analog and hybrid (combined analog/digital) techniques find many uses in the chemical process industries (Distefano, 1968), in environmental modelling (Bullin, 1974) and in biomedical research (Milsum, 1966; Snyder, 1969), it is well worthwhile to review their application to microbial processes.

In the first instance it is important to distinguish between the modes of operation of the analog and the digital computer. As the name implies, an analog computer solves a particular problem if an electrical circuit (or electrical analog) can be set up which obeys the same set of mathematical equations as the problem under study. The electrical circuit is built up from components capable of addition, subtraction, multiplication, division, integration, etc., and once the original problem has been correctly scaled, the output from the circuit will correspond to the dynamic behavior of the original system.

The analog computer is concerned essentially with solving problems of a dynamic nature (i.e., described by differential equations), and as a result of the parallel operation of a large number of electrical components, it provides the solution to a complicated set of equations in negligible solution time. Furthermore, the solution time does not increase significantly when the number of differential equations becomes very large.

The digital computer, on the other hand, operates with discrete numbers which are processed at high speed according to the rules of arithmetic and formal logic. Differential equations are solved by taking small discrete time intervals, and computing the change in the variables which occur over these short time periods. As discussed in the reviews on modelling and simulation, the advent of simulation languages such as MIMIC, DYNAMO, CSMP, etc., has greatly increased the facility of digital computers for solving large numbers of simultaneous differential equations. In contrast to the analog computer, increasing the complexity of a problem appreciably increases solution time.

Numerous applications have been attributed traditionally to the analog computer:

1. Large and complicated sets of differential equations yield fairly readily to analog computation. According to Korn and Korn (1972), modern analog/hybrid computers compete successfully with digital simulation where their inherent speed is really important, i.e., simulation problems requiring thousands of differential equation-solving runs for parameter optimization, contingency studies and statistical evaluations.
2. The analog computer can represent the physical elements of a process. Each block of the computer circuit corresponds to a physical part of the plant, or piece of equipment, and the outputs from the various components of the analog circuit give considerable insight into the start-up and operating characteristics of a process (Rosenbrock and Storey, 1966). A similar application is described by Higgins (1965) with a special analog computer designed to simulate the behavior of various chemical and biochemical reaction networks.
3. An analog computer can be used as a "screening simulator" with various models tested against experimental data in order to build up rapidly a model for more detailed study (Young and Bungay, 1973). The advantage of analog simulation is that it provides immediate interaction with the user through an oscilloscope or $X-Y$ plotter.
4. As an educational facility, analog computers have the advantage of rapid interactive feedback with the user, who can also better visualize the process. Educational programs have been developed in both biochemical engineering and chemical kinetics based on low cost analog/parallel logic computers (Rogers, 1972; Selinger, 1973).

The limitations as well as the possible applications of the analog system must also be considered. Component accuracies for the individual analog-computing elements vary between 0.5 and 0.005% of half-scale (usually 10 volt or 100 volt full-scale) and as Korn and Korn (1972) point out, computation costs increase rapidly if component accuracies higher than 0.1% are desired. By comparison, digital machines compute to high accuracy at relatively low cost. Analog machines do not possess the memory facility of digital computers (for storing data, subroutines, etc.) although it is possible by the use of parallel logic to pre-program the analog computer to carry out sequential computations. Easier programming languages and techniques are being incorporated into the use of both types of computers although the programming of analog computers involves some additional skill in time and magnitude scaling. Table 1 (Röpke and Riemann, 1969) sets out a comparison of the two modes of computation.

Table 1. Comparison between analog and digital computation

Characteristics	Computer type	
	Analog	Digital
Structural elements	Amplifiers, potentiometers, function generators, etc.	Electronic computational and logic elements, magnetic core memory
Working order	Parallel	Sequential
Working time	Not problem-dependent	Problem-dependent
Accuracy	Limited by accuracy of components	High degree of accuracy
Storage capacity	Small–limited to parallel logic (e.g., track/store)	On-line: large Off-line: virtually unlimited
Programming	Relatively simple except for large programs	Facility needed with computer language
Mode of input	Voltage input, direct current operation	Cards, magnetic and paper tape, magnetic disk
Mode of output	*X-Y* plotter, oscilloscope	Line-printer, teletype
Range of applications	Simulation of differential equations and model behavior	Complex computations (including simulation) involving logical operations, data storage, subroutines, etc.

Biochemical engineering applications have been concerned largely with dynamic simulations and with the use of the analog computer to develop kinetic models from experimental data. The need has not yet arisen for the rapid solution of a large number of differential equations, although it is possible that on-line model identification (Shu, 1972) and fermentation optimization could use combined analog/digital computer techniques. As pointed out by Pring (1969), the most desirable machine for fitting a model to the experimental results derived from a complex biological or microbial system is an analog-digital hybrid which combines "the storage capacity and logic of a digital machine with the parallel operation and speed of integration of an analog computer."

2. Characteristics of Analog Computation

2.1 Description

An analog/parallel logic computer which is typical of those found in an electrical engineering laboratory or in chemical process industries is illustrated in Fig. 1. In Section 7 a smaller computer used for educational purposes and capable of modelling most fermentation kinetics problems is discussed.

Fig. 1. Analog/Parallel Logic Computer (courtesy of EAI)

Analog computers differ in a number of ways:

a) capacity – the number of computing components,
b) capability – the quality of the computing components and the operations which they perform,
c) reference voltage level – the operating voltage range of the computer (± 10 or ± 100 volts are typical),

d) factors such as the mode of operation, removability of patch panels, interfacing to output devices, etc.

Many analog computers are equipped with patch panels with input, output and control terminations for the various analog components. Input and output terminations are connected or "patched" together in a configuration defined by the particular problem. The control terminations provide flexibility and allow a component such as an operational amplifier to be used for different mathematical operations.

2.2 Analog Components

The general-purpose analog computer is an assembly of electronic and electromechanical components using DC voltages as variables. Each component can be classified as either linear or non-linear. The linear components (attenuators, operational amplifiers) carry out the following operations:

multiplication by a constant
inversion (change of sign)
algebraic summation
continuous integration.

Non-linear components are used for:

multiplication/division of variables
generation of arbitrary functions (i.e., optimal profiles)
provision of analog memory and elementary logic operations.

Figure 2 illustrates the characteristic symbols for the components. For a fuller description of the detailed operation of specific components, see The Handbook of Analog Computation (EAI, 1971).

As the analog solution to a problem (usually a set of non-linear differential equations) is in the form of a time-varying voltage, output equipment is needed to indicate and record this relationship. A voltmeter or digital panel meter (DPM) provides a simple indicating device while a strip chart or $X-Y$ recorder is suited to continuous recording of output. A cathode ray oscilloscope (CRO) display is often an integral part of an analog computer facility. By programming the computer for rapid, repetitive solution, the output signal can be observed as a continuous visible trace on the screen of the oscilloscope. Rapid interactive feedback between the user and the problem then becomes possible, as any parameter changes can be observed directly.

2.3 Analog Computer Programming

Developing a program for a general-purpose computer can be considered in five sequential steps:

1. Setting up an unscaled circuit diagram
2. Scaling the variables
3. Assigning components
4. Documenting the procedure
5. Carrying out a static check of the circuit

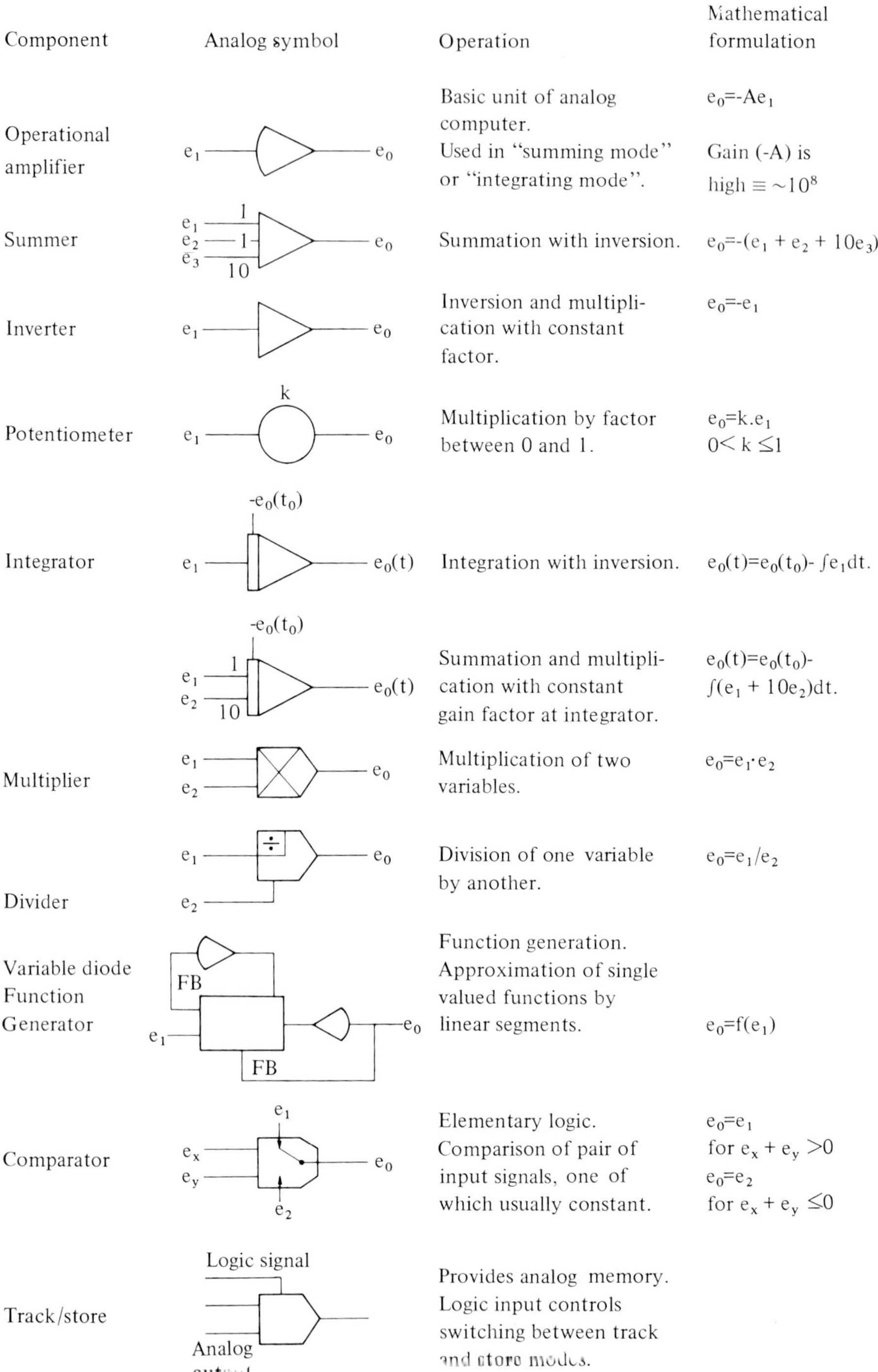

Component	Analog symbol	Operation	Mathematical formulation
Operational amplifier	e_1, e_0	Basic unit of analog computer. Used in "summing mode" or "integrating mode".	$e_0 = -Ae_1$ Gain (-A) is high $\equiv \sim 10^8$
Summer	e_1, e_2, e_3; 1, 1, 10; e_0	Summation with inversion.	$e_0 = -(e_1 + e_2 + 10e_3)$
Inverter	e_1, e_0	Inversion and multiplication with constant factor.	$e_0 = -e_1$
Potentiometer	k; e_1, e_0	Multiplication by factor between 0 and 1.	$e_0 = k.e_1$ $0 < k \leq 1$
Integrator	$-e_0(t_0)$; e_1, $e_0(t)$	Integration with inversion.	$e_0(t) = e_0(t_0) - \int e_1 dt.$
	$-e_0(t_0)$; e_1, e_2; 1, 10; $e_0(t)$	Summation and multiplication with constant gain factor at integrator.	$e_0(t) = e_0(t_0) - \int (e_1 + 10e_2)dt.$
Multiplier	e_1, e_2, e_0	Multiplication of two variables.	$e_0 = e_1 \cdot e_2$
Divider	e_1, e_2, e_0	Division of one variable by another.	$e_0 = e_1/e_2$
Variable diode Function Generator	FB; FB; e_1, e_0	Function generation. Approximation of single valued functions by linear segments.	$e_0 = f(e_1)$
Comparator	e_1; e_x, e_y; e_2; e_0	Elementary logic. Comparison of pair of input signals, one of which usually constant.	$e_0 = e_1$ for $e_x + e_y > 0$ $e_0 = e_2$ for $e_x + e_y \leq 0$
Track/store	Logic signal; Analog output	Provides analog memory. Logic input controls switching between track and store modes.	

Fig. 2. Component elements for analog computer

Unscaled Circuit Diagrams

The general method of programming for high-order differential equations is well documented by Levine (1964) and Fan *et al.* (1970). The equations encountered in studies of fermentation kinetic models are invariably first order, and only first order differential equations will be considered in this review.

Figure 3 shows the unscaled analog circuit for exponential growth, together with the plotted output:

$$\frac{dX}{dt} = \mu \cdot X. \tag{1}$$

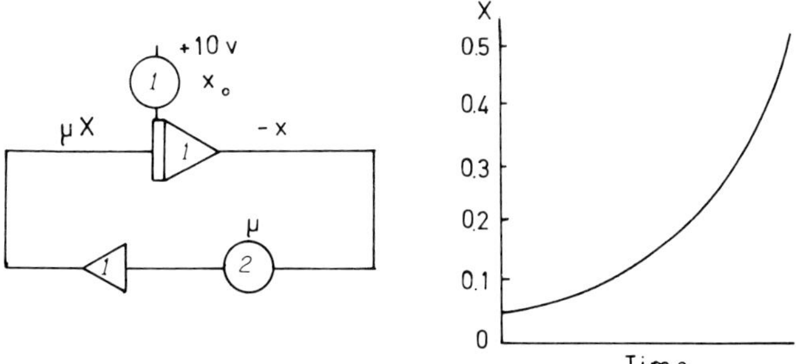

Fig. 3. Analog circuit for exponential growth

In developing the circuit, it is assumed that dX/dt is available to the integrator 1. Output from the integrator is the variable $(-X)$, with the sign change resulting from the integration step. After multiplication by μ (potentiometer 2) and sign inversion (inverter 1), the quantity μX is generated. As dX/dt has now also been generated [see Eq. (1)], the loop is closed and the solution X taken from the inverter output. The problem must be initialized, and the initial conditions ($X_0 = 0.05$) is set on potentiometer 1.

In Fig. 4, the unscaled circuit for the Monod growth model is illustrated:

$$\frac{dX}{dt} = \mu_m \cdot \left(\frac{S}{K_s + S}\right) \cdot X, \tag{2}$$

$$-\frac{dS}{dt} = \frac{\mu_m}{Y_G} \cdot \left(\frac{S}{K_s + S}\right) \cdot X. \tag{3}$$

The initial conditions ($X_0 = 0.05$) and ($S_0 = 1.0$) are set on potentiometers 1 and 4, respectively, and two integrators are required since two first-order differential equations are being solved. Further, two non-linear components, a multiplier and a divider, are necessary as both X and S are variables.

Scaling

Analog computers represent the problem variables by voltages, and dimensional scale factors are chosen so that the values of the variables do not exceed the maximum oper-

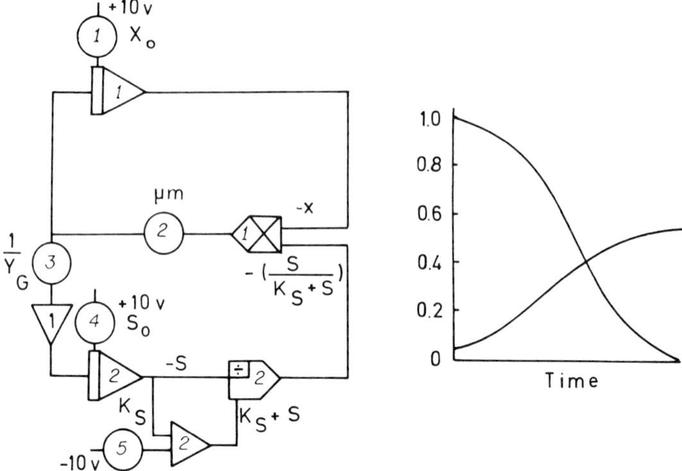

Fig. 4. Analog circuit for Monod growth model

ating voltage of the computer. Scaling can be considered in much the same way as selecting a scale for graphically plotting experimental data.

A magnitude scale factor, κ, is definded as:

$$\kappa = \frac{\text{Reference Voltage}}{\text{Maximum Value of Problem Variable}}.$$

The maximum values of the variables must be estimated for each problem from the available experimental data or from an evaluation of the upper and lower limits within which the system operates.

Once the magnitude scale factor has been estimated, scaled equations are developed by substituting a new scaled variable (e.g., X') for the original problem variable (X) where

$$X' = \kappa \cdot X.$$

Scaling in volts is often practised although the method involves a change in scale factors when the problem is solved on different voltage analog computers. The process of scaling is simplified if the reference voltage of the computer is used as the unit for measuring amplifier output. The method is termed "unit scaling"; i.e., on a 10 volt computer (EAI–180, EAI TR–48), one unit is defined equal to 10 volts. When measured in this convention, all signals will be less than or equal to 1, irrespective of the reference voltage of the computer.

Time scaling is necessary as well as magnitude scaling. The constant relating the time of the computer solution to the actual time is defined as:

$$\tau = \frac{\text{Computer Time}}{\text{Problem Time}}.$$

The time scale factor is chosen to make the potentiometer settings and the gains at the integrators reasonable. All components of an analog computer other than the integrators are unaffected by time scaling. If $\tau > 1$, the solution is slower than the original process; if $\tau < 1$, the computer solution is faster.

For easier scaling, is is recommended that a problem be first programmed and magnitude-scaled without regard to time-scaling. Then by rewriting the scaled equations for the integrators to include the factor $(1/\tau)$, time scaling is effected.

As an example of scaling, the procedure is reported in full for the modelling of penicillin fermentation kinetics (see Section 5.1). Most microbial systems are modelled in terms of first order equations and scaling is relatively simple. For more complex problems, an APSE Compiler (Automatic Programming and Scaling of Equations) written in FORTRAN IV is available for medium and large scale analog computers (EAI, 1973).

Assignment of Components

To facilitate programming and "patching" of the analog computer, each component (amplifier, potentiometer, multiplier, etc.) is assigned a number of address consistent with the numbering system of the computer.

Documentation

Documentation of the procedure used to develop the analog circuit (i.e., original equations, parameter values, unscaled circuit diagram, scaling table, etc.) provides the programmer and operator with sufficient information to check the procedure, make changes in the program and carry out the simulation.

Static Checking

Static checking involves calculating voltages at the output of different components using initial conditions, and comparing these calculated values with actual values measured on the analog circuit. The procedure is advisable in view of the opportunities for error while programming, scaling and "patching" the analog circuit.

3. Simulation of Enzyme Kinetics

As illustrations of analog simulation of the dynamic nature of enzyme reaction kinetics, two familiar examples will be discussed.

3.1 Michaelis-Menten Kinetics

One of the most widely used models of enzyme kinetics is that proposed by Michaelis-Menten. The reaction between enzyme (E) and substrate (S) to form product (P) can be represented as follows:

$$E + S \underset{k_2}{\overset{k_1}{\rightleftharpoons}} E - S \overset{k_3}{\rightarrow} E + P,$$

$$\frac{d[E]}{dt} = -k_1 \cdot [E] \cdot [S] + k_2 \cdot [E-S] + k_3 \cdot [E-S], \tag{4}$$

$$\frac{d[S]}{dt} = -k_1 \cdot [E] \cdot [S] + k_2 \cdot [E-S], \tag{5}$$

$$\frac{d[E-S]}{dt} = k_1 \cdot [E] \cdot [S] - k_2 \cdot [E-S] - k_3 \cdot [E-S], \tag{6}$$

$$\frac{d[P]}{dt} = k_3 \cdot [E-S]. \tag{7}$$

The analog computer solution for the initial conditions S_0 (initial substrate concentration) = 10 g/l, E_0 (initial enzyme concentration) = 2 g/l, and the parameter values $k_1 = 0.50\ \text{sec}^{-1}$, $k_2 = 0.20\ \text{sec}^{-1}$, $k_3 = 0.80\ \text{sec}^{-1}$ is shown in Fig. 5, with the behavior of product, substrate, enzyme and enzyme-substrate complex clearly characterized.

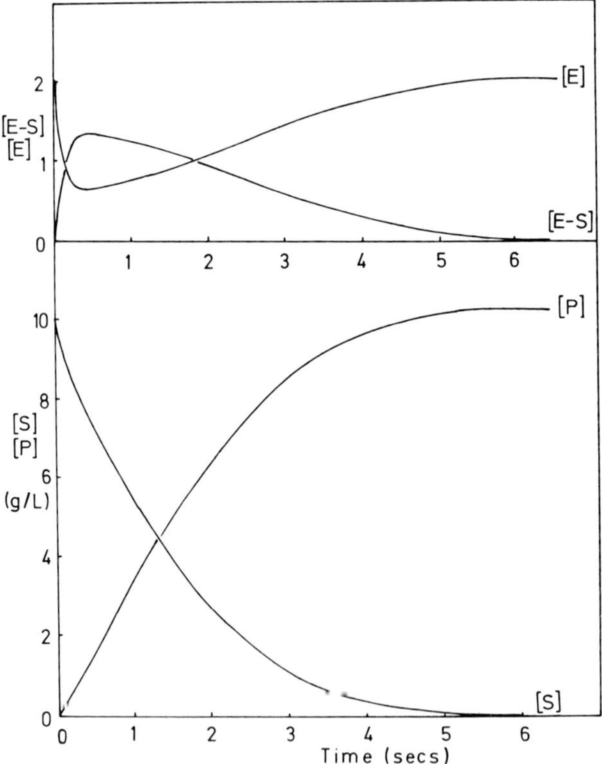

Fig. 5. Simulation of Michaelis-Menten kinetics

3.2 Competitive Enzyme Inhibition

Competitive enzyme inhibition is described by the following equations:

$$E + S \underset{k_2}{\overset{k_1}{\rightleftharpoons}} E - S \overset{k_3}{\rightarrow} E + P,$$

$$E + I \underset{k_5}{\overset{k_4}{\rightleftharpoons}} E - I.$$

It is assumed that two different equilibria are established in the above sequence of reactions—one between enzyme, substrate and enzyme-substrate complex, and the other between enzyme, inhibitor and enzyme-inhibitor complex.

In a similar kinetic analysis to that involved in Michaelis-Menten kinetics, differential equations were developed to describe the rate of change of each component or complex. Initial conditions $S_0 = 10$ g/l and $E_0 = 2$ g/l were selected, together with parameter values $k_1 = 0.50\ \text{sec}^{-1}$, $k_2 = 0.20\ \text{sec}^{-1}$, $k_3 = 0.80\ \text{sec}^{-1}$, $k_4 = 0.10\ \text{sec}^{-1}$, and $k_5 = 0.05\ \text{sec}^{-1}$. By increasing the initial concentration of the competitive inhibitor (I_0) from 2 g/l to 10 g/l, a sensitivity analysis was carried out (Fig. 6) illustrating the effect of increasing I_0 on the rate of product formation.

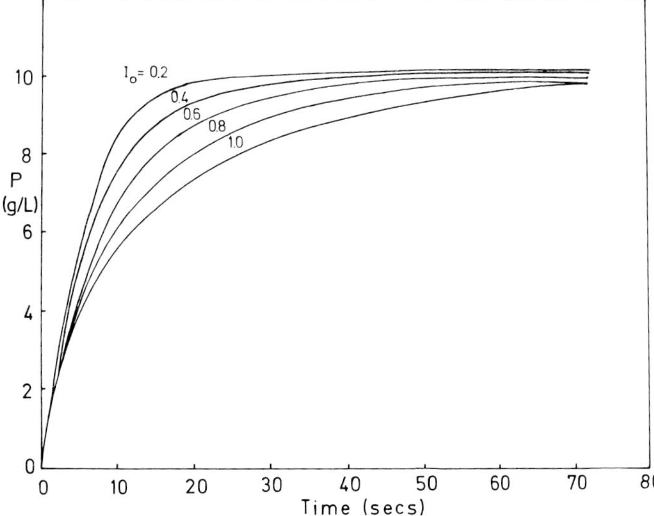

Fig. 6. Sensitivity analysis for competitive enzyme inhibition

4. Analog Modelling of Cellular Control Process

Analog computer simulations have been used to investigate the models for cellular control processes proposed by Goodwin (1963, 1965), and to relate experimentally observed cytochrome oscillations (Gray and Rogers, 1971a,b) to the theoretical models. The models consider that significant biosynthetic pathways within the cell are functioning under conditions of feedback control. The different components of the pathway are

1. a genetic locus at which mRNA is synthesized,
2. a ribosomal site for enzyme synthesis, and
3. a metabolic site at which the enzyme interacts with a substrate to bring about product formation.

Differential equations proposed by Goodwin predict that for certain parameter values, the concentrations of mRNA, enzyme and metabolic product follow oscillatory patterns after a disturbance. Metabolic disturbances have been imposed on different experimental systems and oscillations reported in the growth rate of a culture (Dean and Rogers, 1967; Dean and Moss, 1971) and in particular metabolites (Knorre, 1968; Pye, 1969; Harrison, 1970). A recent review of oscillatory states in continuous culture (Harrison and Topiwala, 1974) discusses some of the available evidence for the existence of oscillations derived from intracellular feedback regulation.

In the *study of cytochrome oscillations,* chloramphenicol was used to interrupt A- and B-group cytochrome synthesis of *Candida utilis* growing in a chemostat. The growth rate was unaffected by chloramphenicol addition, and recovery and oscillations in the cytochrome levels were observed when the addition of chloramphenicol was terminated. In Fig. 7 the pattern of damped oscillations in A-group cytochromes is illustrated, and

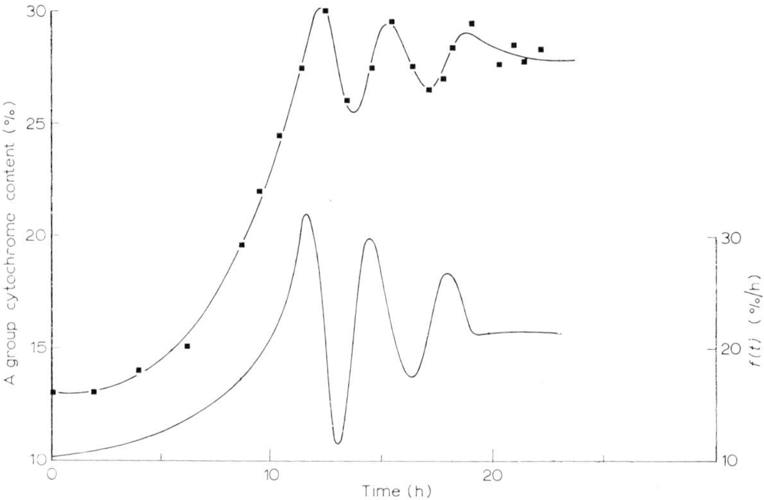

Fig. 7. A-group cytochrome content of *C. utilis* ($D = 0.3\ h^{-1}$) following washout of 2.0 g/l chloramphenicol fitted by Fourier analysis. Lower curve is the rate of synthesis of A-group cytochrome (after Gray and Rogers, 1971)

similar oscillations at different frequencies were reported for other growth conditions. Based on the conceptual models proposed by Goodwin, two particular models were investigated using analog simulation:

1. a cytochrome feedback repression model in which mRNA repression results from the increasing concentration of cytochromes,
2. a product feedback repression model in which mRNA repression is caused by the action of an associated metabolic product. The constants in the model were based on experimental determination and on values reported by Goodwin for rates of protein synthesis.

In the analog simulation of chloramphenicol addition, a step input was imposed by switching from one potentiometer to another, the latter corresponding to the steady state drug concentration in the chemostat. Washout was simulated by switching back to the original position, although in the experimental system some time would elapse before the drug concentration returned to zero.

From Fig. 8 it is evident that the model involving cytochrome feedback repression does not exhibit oscillatory behavior. The model which assumes that a metabolic product other than the cytochromes represses mRNA synthesis was found to be capable of predicting an oscillatory response. Further, decreased oscillation period with increased dilution rate indicated by the model was in qualitative agreement with empirical observations.

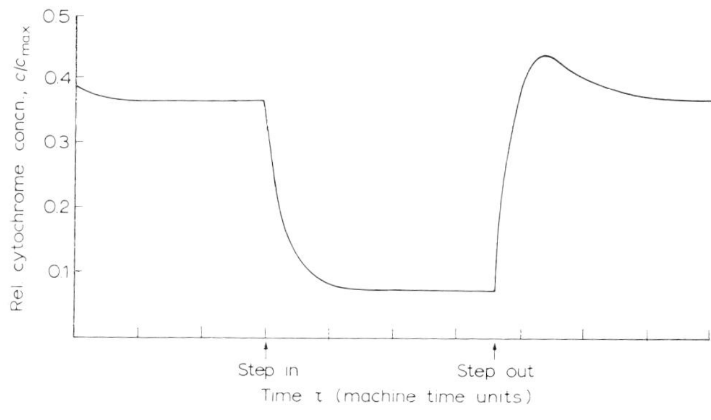

Fig. 8. (a) Analog simulation for model involving cytochrome repression of mRNA synthesis.

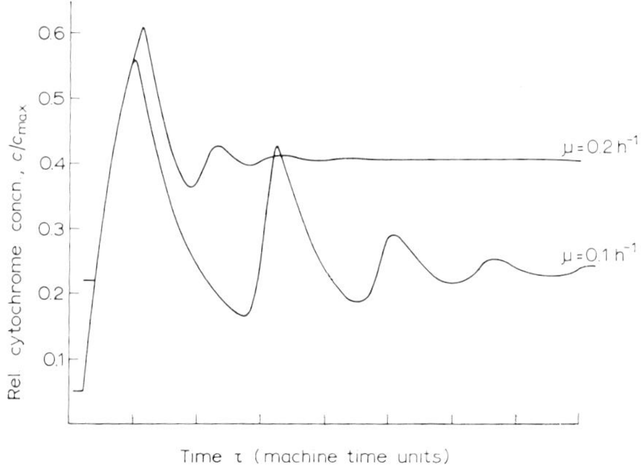

Fig. 8. (b) Analog simulation for model postulating that a metabolic product is involved in repression of mRNA synthesis (after Gray and Rogers, 1971)

5. Analog Modelling of Fermentation Kinetics

5.1 Penicillin Fermentation

As an illustration of the scaling of the analog computer to typical fermentation processes, the data published by Constantinides *et al.* (1970) for a penicillin fermentation is considered, together with the models proposed by the above authors for cell growth and product synthesis. The kinetic equations are as follows:

$$\frac{dX}{dt} = b_1 \cdot X \cdot (1.0 - \frac{X}{b_2}), \tag{8}$$

$$\frac{dP}{dt} = b_3 \cdot X - b_4 \cdot P. \tag{9}$$

The first equation takes the form of a logistics law where b_1 and b_2 are empirical constants. It can be considered to include a first-order birth term and a second-order death term. The second equation describes the rate of synthesis of penicillin, where the rate of synthesis is proportional to the cell concentration. The second term in Eq. 9 represents a first-order penicillin destruction mechanism (b_3 and b_4 are empirical constants).

Scaling

Variable	Estimated Max. Value	Scale Factor	Scaled Variables
X	50 g/kg	0.02	$(0.02X)$
P	10000 units/ml	0.0001	$(0.0001\,P)$

Magnitude scaling is based on the convention of "unit scaling", i.e., making the reference voltage equal to 1 unit.
Defining $X' = 0.02X$ and $P' = 0.0001P$, the equations become:

$$\frac{dX'}{dt} = b_1 \cdot X' \cdot (1.0 - \frac{X'}{(0.02b_2)}), \tag{10}$$

$$\frac{dP'}{dt} = (0.005b_3) \cdot X' - b_4 \cdot P'. \tag{11}$$

The time-scaled factor τ can be calculated for a solution time of 10000 sec and for an actual fermentation time of 250 hrs. The solution time is chosen so that the coefficient values set on the potentiometers are "reasonable"; a good general rule is that they should be between 0.1 and 5.0.

$$\tau = \frac{10000}{250 \times 3600} = \frac{1}{90}.$$

Defining $T = \tau \cdot t$, Eqs. (10) and (11) can be rewritten:

$$\frac{dX'}{dT} = (90b_1) \cdot X' \cdot (1.0 - \frac{X'}{(0.02b_2)}) \tag{12}$$

$$\frac{dP'}{dT} = (0.45b_3) \cdot X' - (90b_4) \cdot P' \tag{13}$$

Parameter and Potentiometer Values

Pot. No.	Parameter	Scaled parameter	Pot value
1	$(X)_0$	$(X')_0$	0.01
2	b_1	$90b_1$	0.369 (Gain 10)
3	$1/b_2$	$1/(0.02b_2)$	0.115 (Gain 10)
4	b_3	$0.45b_3$	0.417 (Gain 10)
5	$(P)_0$	$(P')_0$	0
6	b_4	$90b_4$	0.369 (Gain 10)

where $b_1 = 0.041$; $b_2 = 43.3$; $b_3 = 9.27$; $b_4 = 0.041$ at 22° C. It should be noted that a solution time of 10000 sec was selected to give a suitable value of τ and the scaled parameters. In practice a solution time of 100 sec. is more suitable and can be readily achieved by decreasing the solution time 100-fold on the computer.

The analog circuit and the simulation curves for the model describing penicillin production are shown in Figs. 9 and 10.

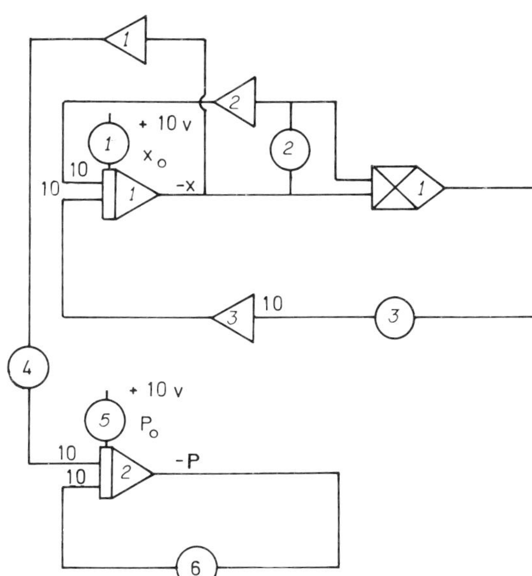

Fig. 9. Scaled analog circuit for modelling penicillin fermentation kinetics

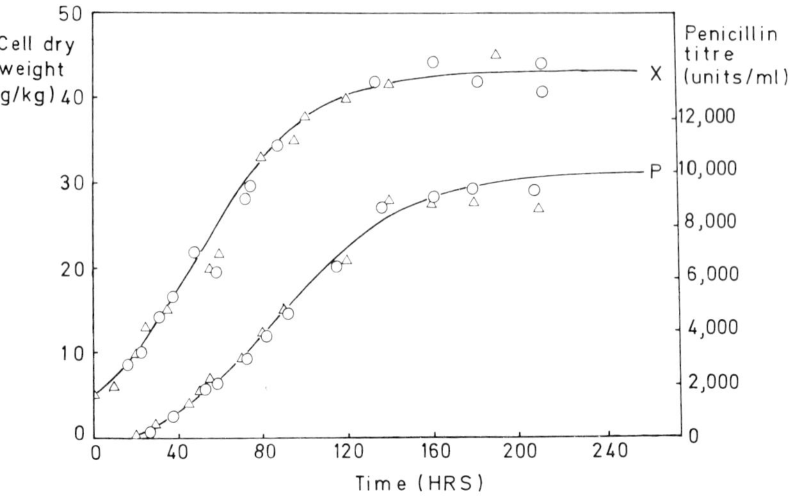

Fig. 10. Analog simulation of penicillin fermentation kinetics (data from Constantinides *et al.*, 1970)

5.2 Microbial Leaching of Minerals

In the leaching of low-grade ores such as the sulphides of copper, iron, nickel, zinc, etc., and the oxides of uranium, both chemical and microbiological processes are active (Corrans *et al.*, 1972; Tuovinen and Kelly, 1972). The development of a realistic model for microbial leaching depends on testing a number of model structures and parameter values against empirical data, and in the early stages of formulation, analog simulations can prove invaluable.

From a study of the microbial leaching of a synthetic nickel sulphide (NiS) by *Thiobacillus ferrooxidans*, a kinetic model has been proposed (Rogers and Hendy, 1976) which considers:

1. direct bacterial action,
2. indirect bacterial action, and
3. acid leaching of NiS

The model for the leaching of synthetic NiS (a mineral with a 1 : 1 stoichiometric relationship between Ni and S) is based on the following reactions:

$$NiS + 2\,Fe^{3+} \xrightarrow{k_1'} Ni^{2+} + S + 2\,Fe^{2+},$$

$$2Fe^{2+} \underset{\text{bacteria}}{\xrightarrow{k_2'}} 2Fe^{3+} + 2e,$$

$$NiS + 2O_2 \underset{\text{bacteria}}{\xrightarrow{k_3'}} Ni^{2+} + SO_4^{2-}$$

$$NiS + H_2SO_4 \xrightarrow[(pH=2.5)]{k'_4} Ni^{2+} + SO_4^{2-} + H_2S\uparrow.$$

Natural oxidation of the NiS and Fe^{2+} will also occur in the absence of bacteria although it has been shown that the rates of natural oxidation are comparatively slow (i.e., for Fe^{2+} the rate of natural oxidation is 10^{-5} to 10^{-6} slower than that occuring in the presence of bacteria). For the oxidation of Fe^{2+} by bacteria, electrons are released and a further reaction could be written for the acceptance of these electrons by oxygen.
Based on the above stoichiometry and assuming growth of bacteria on the available Fe^{2+} rather than on precipitated sulphur (the latter is much slower), kinetic equations can be developed:

$$\frac{d[Ni^{2+}]}{dt} = k'_1 \cdot [Fe^{3+}] + k'_3 \cdot X + k'_4 \ , \tag{14}$$

$$\frac{d[Fe^{2+}]}{dt} = 2k'_1 \cdot [Fe^{3+}] - k'_2 \cdot \left(\frac{[Fe^{2+}]}{K_s + [Fe^{2+}]}\right) \cdot X, \tag{15}$$

$$\frac{d[Fe^{3+}]}{dt} = -2k'_1 \cdot [Fe^{3+}] + k'_2 \cdot \left(\frac{[Fe^{2+}]}{K_s + [Fe^{2+}]}\right) \cdot X - k'_5 \cdot [Fe^{3+}], \tag{16}$$

$$\frac{dX}{dt} = k'_6 \cdot \left(\frac{[Fe^{2+}]}{K_s + [Fe^{2+}]}\right) \cdot X. \tag{17}$$

It has been established that the rate of NiS leaching by Fe^{3+} is first order, and that the ratio of Ni^{2+} solubilized to Fe^{3+} reduced is 1 : 2. The total iron in the system decreased with time (over 200 hrs) and a first order precipitation term was included in Eq. (16). The rate of Fe^{2+} oxidation by bacteria was assumed to follow a Monod relationship, and k'_2 was related to k'_6 by $k'_2 = k'_6/Y$ where Y was the yield of cells/ppm Fe^{2+}. The cell concentration (X) has not been accurately determined in the experimental system due to the problem of estimating cells when a high proportion are attached to a mineral surface.
Several model formulations were tested and in Fig. 11, simulation curves for the Ni^{2+}, Fe^{2+}, and Fe^{3+} are compared against the experimental values. Good agreement is observed for the parameter values specified. Sensitivity testing was carried out for the different parameters and in Fig. 12, the effect of varying K_S in the range 500 – 1500 ppm Fe^{++} is shown. These values of K_s are comparable with those proposed by Lacey and Lawson (1970) for *T. ferrooxidans*. A further effect which could be included in the model is the inhibition of growth by high levels of ferric iron (Wong *et al.*, 1974). However, as shown in Fig. 11, the level of Fe^{3+} fell fairly rapidly to 150 – 200 ppm, and the inhibition effect would be small.

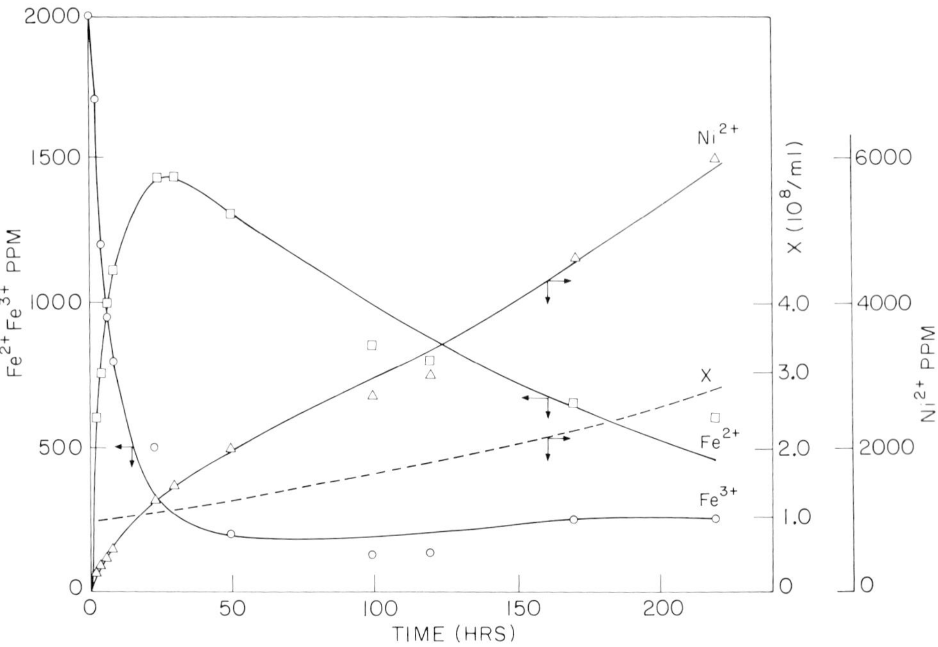

Fig. 11. Analog simulation of microbial leaching of synthetic nickel sulphide (NiS) by *T. ferrooxidans* Parameter Values: $k_1' = 0.060$ ppm ppm $(Fe^{3+})^{-1}h^{-1}$; $k_2' = 40$ ppm 10^8 cell$^{-1}h^{-1}$ml; $k_3' = 0$; $k_4' = 12$ ppm h^{-1}; $k_5' = 0.025$ ppm ppm $(Fe^{3+})^{-1}h^{-1}$; $k_6' = 0.012\ h^{-1}$; $K_S = 1300$ ppm (Fe^{2+})

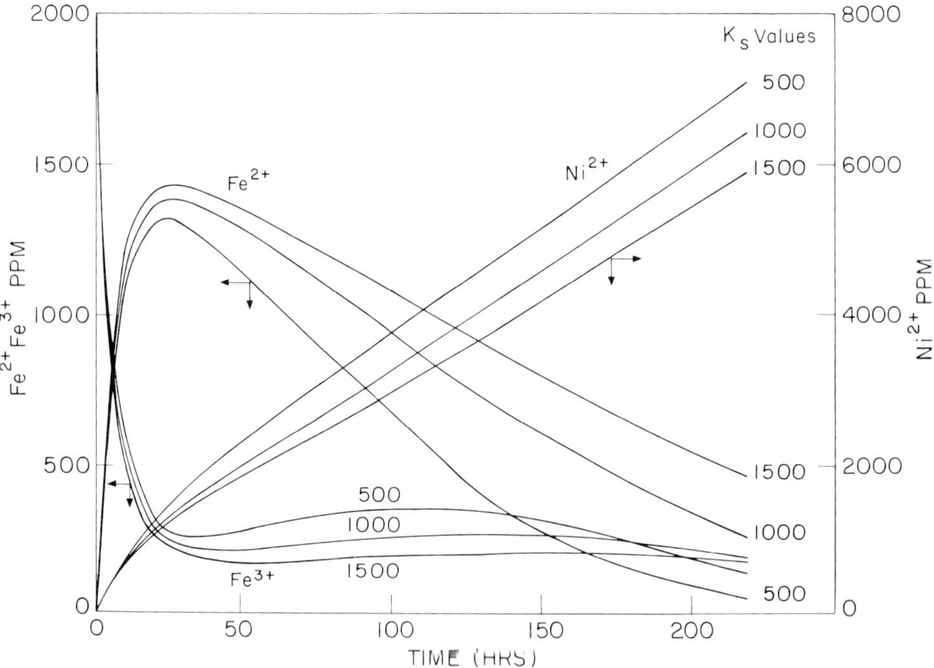

Fig. 12. Sensitivity analysis of model for microbial leaching of NiS. The value of K_S is changed from 500 ppm to 1500 ppm, with all other parameters constant

6. Analog Simulation of Continuous Processes

Analog computation is concerned essentially with solving problems of a dynamic nature, and as such is ideally suited to studying transient behavior of systems. In fact, it is not uncommon to determine parameter values from experimental batch data using least square regression analysis and for the dynamic behavior of the fermentation during "start-up" to be simulated on an analog computer (Blanch and Rogers, 1971; Hanson and Tsao, 1972).

Chemostat transient behavior based on Monod kinetics can be modelled:

$$\frac{dX}{dt} = \mu_m \cdot \left(\frac{S}{K_S + S}\right) \cdot X - D \cdot X, \tag{18}$$

$$\frac{dS}{dt} = D \cdot (S_0 - S) - \frac{\mu_m}{Y_G} \cdot \left(\frac{S}{K_S + S}\right) \cdot X. \tag{19}$$

However, the simple Monod model has many limitations when predicting the transient response of a microbial culture, and simulation studies which assume a first order lag in growth rate response to changes in substrate concentration (Fig. 13) have shown more

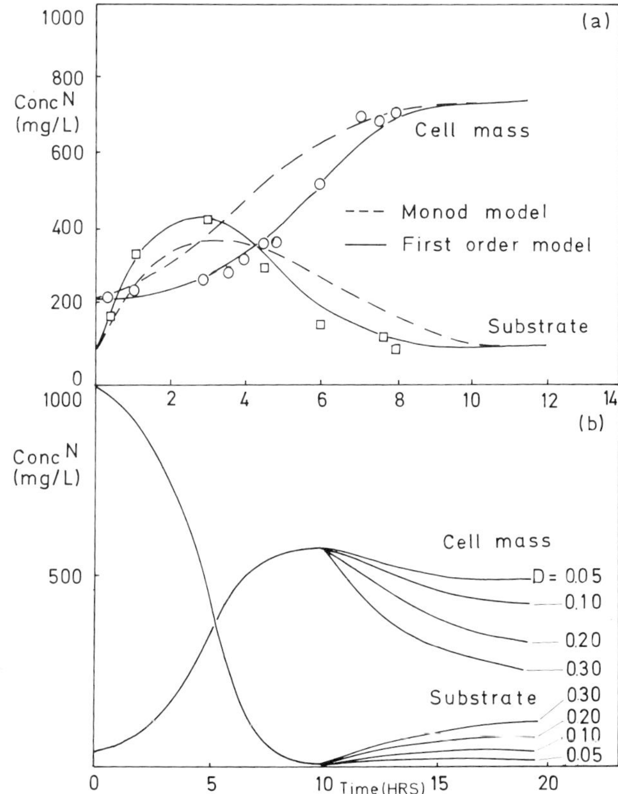

Fig. 13. (a) Transient analysis of continuous culture using analog simulation. The Monod model and a first-order response model are shown (data from Young *et al.*, 1970). (b) The effect of dilution rate (D) on continuous culture "start-up" based on the Monod model

realistic behavior (Young *et al.*, Young and Bungay, 1973). Further simulations have considered the effect of variable yield and the influence of growth inhibitors (Yano and Koga, 1969; Zines and Rogers, 1970).

Biological waste treatment processes have been modelled in a similar manner to the chemostat, on the assumptions of growth limitation and inhibition by a single substrate and the production of uniform "active" biomass. Based on analog simulations, recycling biomass and preconcentration of nutrients were shown to be critical factors in waste treatment (Andrews, 1968, 1971). A further important factor–the cycling nature of the input stream was simulated (Fan *et al.*, 1970) and it was predicted that performance could be influenced significantly by input frequencies close to diurnal fluctuations. The effect of mixed culture interactions have also been given some attention (Bungay and Bungay, 1968) in an attempt to develop a model structure for the multiple substrate limitations, multiple inhibitions and mixed culture interactions which constitute a practical waste treatment process.

A simple model for interaction between two species can be written in terms of the familiar predator/prey (or host/parasite) relationship. For a continuous process, the equations become:

$$\frac{dH}{dt} = \mu \cdot H - k_1'' \cdot H \cdot P_r - D \cdot H, \tag{20}$$

$$\frac{dP_r}{dt} = k_2'' \cdot H \cdot P_r - D \cdot P_r. \tag{21}$$

It has been assumed that no substrate limitation exists for the growth of the hosts (H), and as shown in Fig. 14 a, a pattern of oscillations in both H and P_r can be generated by the model (for a fuller discussion of the interactions, see Canale 1969, 1970). Figure 14 b illustrates the effect of increasing dilution rate on the amplitude of the oscillations.

Several other interesting analog simulations of continuous processes include the use of simulation to design an optimal membrane area for dialysis of fermentation products (Schultz and Gerhardt, 1969) and modelling a continuous enzymatic reactor with an associated ultrafiltration unit (Bowski *et al.*, 1971, 1972). In the latter case the model was developed from empirical data, while in both examples a range of parameter values were tested for their effect on system response.

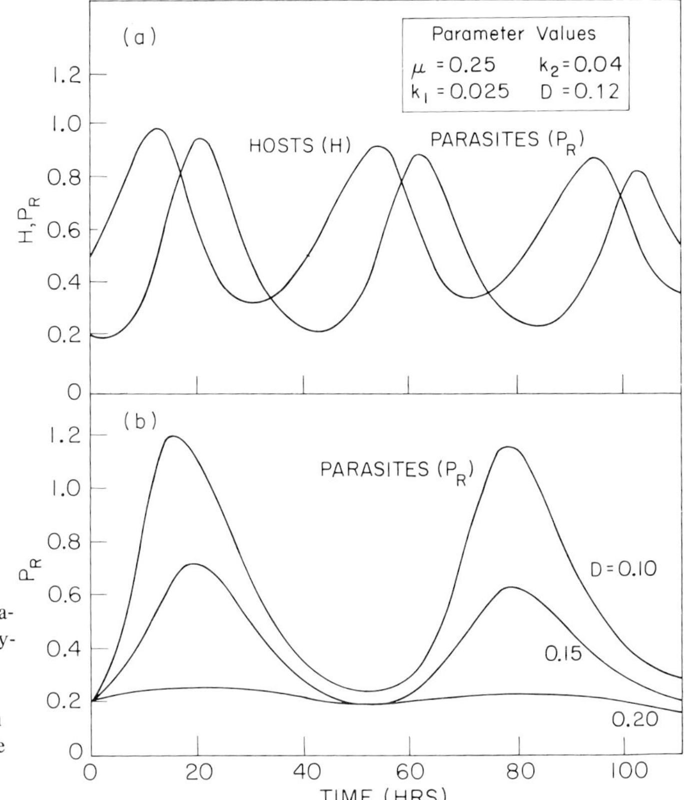

Fig. 14. (a) Analog simulation of a simple predatory-prey model for mixed culture interactions. (b) The effect of dilution rate (D) on the amplitude of mixed culture interactions

7. The Educational Role of the Analog Computer

The analog computer has several advantages in an educational program as a result of direct interaction between the user and the problem, together with the capability for rapidly testing a range of model types and parameter values. In simulation studies and model building where a high level of accuracy is not a prerequisite, a low cost analog computer (Fig. 15) together with a suitable output device (i.e., cathode ray oscilloscope or $X-Y$ plotter) can provide an excellent laboratory facility.

The multidisciplinary nature of biochemical engineering, and the related area of fermentation technology, attracts students with diverse backgrounds. Some have had considerable experience in quantitative methods, computer simulation and the general systems analysis approach to problem solving. Others, concentrating on the more biological and less quantitative approach, find the description of problems in terms of differential equations unfamiliar.

It has been found that the use of a small analog computer programmed to solve problems similar to those outlined in this paper provides one way of bridging the gap between the

Fig. 15. Small low-cost analog computer suitable for educational programs (courtesy of EAI)

different groups of students. The techniques of simple analog programming and scaling are rapidly acquired (well within one laboratory session), allowing the student to proceed quite expeditiously to sensitivity analysis and simulation runs for problems involving 2 or 3 different equations. Rapid feed-back provided by the output device generates significant interaction between the student and the problem at hand. Furthermore, the problems can be developed on a self-paced learning basis, with a number of sequential steps for the student to follow.

Most courses in biochemical engineering attach considerable importance ot the subject areas of enzyme kinetics, fermentation kinetics, continuous processes and enzyme technology, and it is possible to develop simulation examples for each of these areas (Table 2).

Table 2. Simulation of microbial systems with the EAI-180 Analog/Logic Computer

Enzyme kinetics	*Original Reference*
Michaelis-Menten kinetics	Aiba *et al.* (1973)
Competitive enzyme inhibition	Aiba *et al.* (1973)
Enzyme technology	
Sucrose hydrolysis on invertase	Bowski *et al.* (1972)
Fermentation kinetics	
Industrial alcohol production	Aiba *et al.* (1973)
Lactic acid fermentation	Rogers and McDonald (1976)
Penicillin fermentation	Constantinides *et al.* (1970)
Gramicidin S modelling	Blanch and Rogers (1971)
Microbial leaching (NiS)	Rogers and Hendy (1976)
Metabolic oscillations	
Cytochrome oscillations	Gray and Rogers (1971)
Continuous microbial systems	
Dynamic simulation of continuous culture	Young *et al.* (1970)
Stability analysis of chemostat	Koga and Humphrey (1967)
Interacting microbial cultures	Bungay and Bungay (1968)
Models for waste treatment	Andrews (1968, 1971)
Multistage continuous culture	Aiba *et al.* (1973)

As implied earlier, the student who is less familiar with simulation techniques can proceed to a fairly simple analysis of the problem, while the more quantitative student can develop the problem more fully on a larger analog computer or using digital simulation. As well as providing the educational facility for simulation, analog techniques are very useful in allowing students to develop models from laboratory data. Many kinetic formulations of microbial systems are in relatively simple form. The production of lactic acid or alcohol, for example, can be described in terms of "growth associated" and "non-growth associated" kinetics (Luedeking and Piret, 1959; Aiyar and Luedeking, 1966) while rates of enzyme synthesis can be modelled in a similar way (Gray *et al.*, 1972; Terui, 1973). Secondary metabolite production has been modelled with "non-growth associated" kinetics (Section 5.1), although more realistic models depend on stoichiometric mass balances (Calam *et al.*, 1971). As the early stages of model building are essentially heuristic, the analog computer can be used as a "screening simulator" to build a model from simple beginnings and concurrently design a strategy for further experimentation.

No discussion of the educational role of analog computation would be complete without considering the complementary and in some ways overlapping role played by digital computation. In this review, emphasis has been placed on the use of low cost analog computation to study relatively simple kinetic models. All of the problems discussed could be equally well studied using digital computation, although perhaps not quite as expeditiously or instructively. As the complexity of the model increases, the advantages of using a simulation language and digital computation begin to outweigh the benefits of simple analog methods (Blanch and Dunn, 1974). Accurate estimation of parameters from kinetic data is also carried out most effectively on a digital computer although the early stages of model development and experimental design may have resulted from analog studies.

8. Hybrid Computation

The hybrid computer would appear to have significant potential in the on-line analysis of fermentation kinetics and optimization of microbial processes as it combines the advantages of analog and digital computation. The facility for high-speed solution of differential equations and real-time simulation is provided by the analog, while the digital component handles the computational flow, the optimization algorithm and the control strategy for the process. As a result, hybrid applications are frequently found in chemical process industries and in biomedical research.

The first step in optimizing a fermentation involves developing a realistic model. By comparison with a chemical reaction which is defined stoichiometrically, modelling of a microbial process is inherently complex. However, as pointed out in the discussion of model building using analog techniques, most current models of fermentation processes are relatively simplistic and in the long term clearly inadequate. Most are based on Monod and product inhibition kinetics although such relationships only describe the growth of a culture in the decline phase following the achievement (if this ever occurs) of the maximum growth rate (μ_m). Such models are also based on steady state correlations, and are not capable of predicting dynamic changes in a batch culture subject to programmed feeding of nutrients, addition of precursors or changing profiles of temperature, pH, dissolved oxygen, etc. As programmed feeding is practised in a number of industries (Hockenhull and McKenzie, 1968; Reed and Peppler, 1973) and as it has been predicted that changing temperature profiles give increased productivities, more dynamic models are necessary.

Computer analyses will clearly play a major role in this model development. To illustrate the potential of hybrid computation, a generalized kinetic model similar to those used for modelling alcohol or lactic acid production will be considered. The model, based on simple relationships, takes account of substrate limitation and inhibition, product inhibition, cell mortality and maintenance energy requirements, i.e.,

$$\frac{dX}{dt} = \mu_m \cdot \left(\frac{S}{K_S + S}\right) \cdot \left(\frac{K_P}{K_P + P}\right) \cdot \left(\frac{K_I}{K_I + S}\right) \cdot X - \gamma \cdot X, \tag{22}$$

$$-\frac{dS}{dt} = \frac{1}{Y_G} \cdot \left(\frac{dX}{dt}\right) + \frac{1}{Y_P} \cdot \left(\frac{dP}{dt}\right) + m \cdot X, \tag{23}$$

$$\frac{dP}{dt} = \alpha \cdot \left(\frac{dX}{dt}\right) + \beta \cdot X. \tag{24}$$

The model considers 3 main variables and involves 10 parameters without taking account of the correlations which could be developed between the parameters and the control variables. The maximum growth rate (μ_m) will be a function of these variables, for example, as will be the parameters α and β, i.e.,

$$\mu_m = f(T, \text{pH}, DO, \ldots). \tag{25}$$

As shown diagramatically in Fig. 16, the hybrid computer is well suited to kinetic modelling as the number of parameters increases and as a large data set is developed. On-line hybrid analysis of the above model with 10 parameters could be readily programmed. Frequent on-line sampling and data analysis will generate a large data set for parameter estimations. Furthermore, increasing the number of parameters in the model, i.e., by

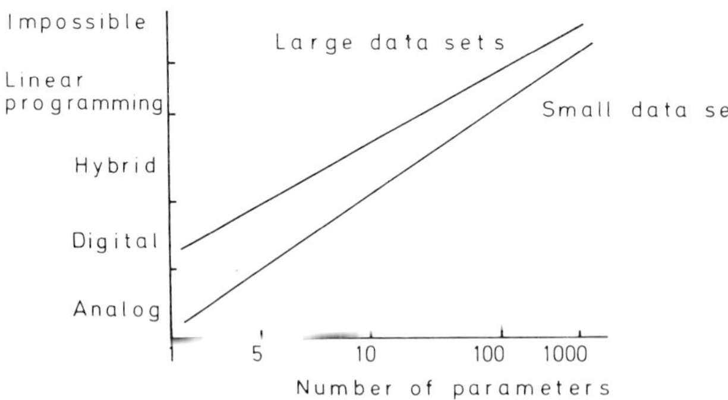

Fig. 16. The effect of increasing the number of parameters and the size of the data set on the selection of the mode of computation (after Distefano, 1968)

incorporating time constants and the effects of control variables, will render the use of hybrid computation even more advantageous. Figure 17 illustrates the highly interactive roles of both analog and digital components in the on-line modelling.

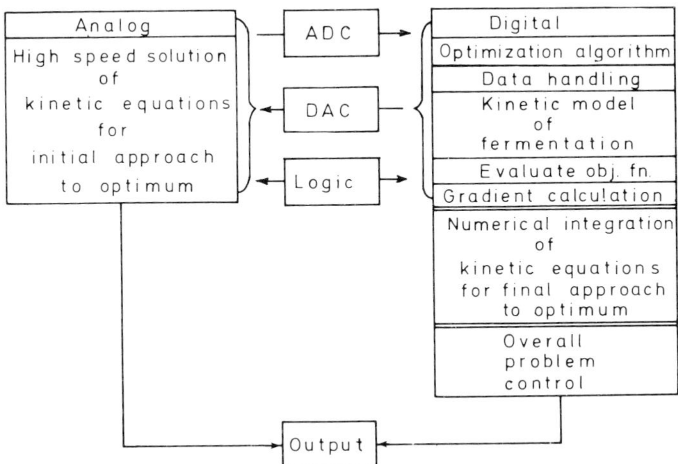

Fig. 17. Interaction between analog and digital techniques in developing a kinetic model of a fermentation process

Once a fermentation has been modelled adequately, computer controlled optimization becomes a real possibility. As discussed by Unden and Heden (1973) and Unden (1974), different optimization techniques are available and range from a purely empirical approach to the methods of dynamic optimization used in the process industries. In the former case the objective function is calculated for different values of the control variables and a search scheme developed to locate the optimum. However, the method does not come to terms with the transient characteristics of a process and is confined to calculations at different steady-state conditions (as occur in continous cultivation). Dynamic optimization recognizes the dynamics associated with a process and manipulates the system inputs (i.e., feed rate, temperature) to continuously maximize an objective function. The technique requires extensive mathematical modelling and significant amounts of on-line computer core memory in order to cope with the calculation requirements (Williams, 1970). In the fermentation industry dynamic optimization would be best suited to batch fermentations with programmed feeding, control variable profiles, etc., and potentially to multistage continuous cultivation.

A possible strategy for dynamic optimization of a fermentation is shown in Fig. 18. The method would be accomplished best by using hybrid computation, and would involve off-line kinetic modelling and optimization in fast time in parallel with the real time operation of the actual process. The optimum settings of the controller of nutrient feed pump would be calculated from an off-line computation involving a model of the controller, a simulation of the fermentation and an optimization algorithm set to maxi-

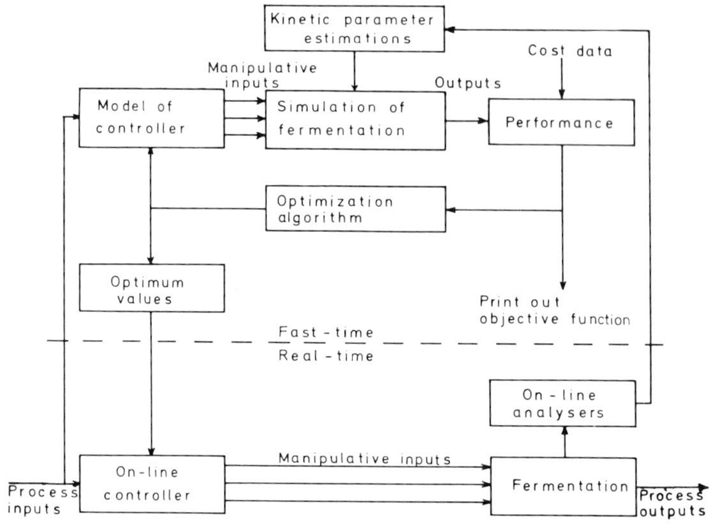

Fig. 18. Strategy for the dynamic optimization of a fermentation process using hybrid computing techniques

mize the yield or productivity of the fermentation. Feedback from the process is provided by on-line automatic analysers, and continuous updating of the kinetic model would be programmed throughout the batch process. The optimum settings of the controller or feed pump would depend on the state of the fermentation and would be likely to change significantly over the course of a production run. It is also important to note that by operating the off-line computation at sufficiently high speed, the total time spent in optimizing the process is negligible compared with the time response of the real process (i.e., in real time).

9. Conclusions

It has been the objective of the review to evaluate analog/hybrid computation and to look at the areas in biochemical engineering where it is most likely to be applied. Clearly there is a vast area where pure analog and digital techniques overlap and in many cases digital solutions are far superior, being simpler and less costly. However, at both ends of the computational spectrum a case can be made for the analog or the hybrid (analog/digital) approach. Simple educational simulations involving two or three differential equations are easily solved on the analog computer and the interactive nature of the problem-solving makes the analog an excellent teaching facility. The early stages of model building bases on a small number of equations, fall into a similar category. At the more distant end of the spectrum, the hybrid computer offers considerable potential in the dynamic optimization of fermentation processes. However, on-line optimization of microbial systems is in its infancy, and time, experience and economic criteria will the final judges.

Nomenclature

Symbols

$b_1 \dots b_4$ = Kinetic constants for penicillin model
D = Dilution rate (h^{-1})
E = Enzyme concentration (g/l)
H = Host concentration (g/l)
I = Inhibitor concentration (g/l)
$k_1 \dots k_5$ = Constants for enzyme kinetics
$k'_1 \dots k'_6$ = Constants for microbial leaching model
k''_1, k''_2 = Constants for host/parasite problem
K_S = Saturation constant (g/l)
K_P = Product inhibition constant (g/l)
K_I = Substrate inhibition constant (g/l)
m = Specific maintenance energy requirements
P = Product concentration (g/l)
P' = Scaled product concentration (g/l)
P_r = Parasite concentration
S = Substrate concentration (g/l)
S_0 = Initial substrate concentration (g/l)
t = Real time
T = Computer time
X = Biomass concentration (g/l)
X' = Scaled biomass concentration (g/l)
Y = Yield from energy source (g biomass/ppm Fe^{2+})
Y_G = Yield for biomass production (g biomass/g substrate)
Y_P = Yield for product formation (g product/g substrate)

Greek letters

α = "Growth associated" kinetic constant
β = "Non-growth associated" kinetic constant
γ = Specific death rate (h^{-1})
κ = Magnitude scale factor
τ = Time scale factor
μ = Specific growth rate (h^{-1})
μ_m = Maximum specific growth rate (h^{-1})

References

Aiba, S., Humphrey, A. E., Millis, N.: Biochemical Engineering, 2nd edition, University of Tokyo Press (1973).

Aiyar, A. S., Luedeking, R.: CEP Symp. Ser. **62**, 55 (1966).

Andrews, J. F.: Biotechnol. Bioeng. **10**, 707 (1968).

Andrews, J. F.: Biotechnol. Bioeng. Symp. Series No. **2**, 5 (1971).

APSE (Automatic Programming and Scaling of Equations), Electronic Associates Inc. (1973).

Blanch, H. W., Dunn, I. J.: In: Advances in Biochemical Engineering 3 (Eds. T. K. Ghose, A. Fiechter, and N. Blakebrough), Springer-Verlag, p. 127 (1974).

Blanch, H. W., Rogers, P. L.: Biotechnol. Bioeng. **13**, 843 (1971).

Bowski, L., Saini, R., Ryu, D. Y., Vieth, W. R.: Biotechnol. Bioeng. **13**, 641 (1971).

Bowski, L., Shah, P. M., Ryu, D. Y., Vieth, W. R.: Biotechnol. Bioeng. Symp. Series No. **3**, 229 (1972).

Bullin, J. A.: Environ. Sci. Tech. **8**, 156 (1974).
Bungay, H. R., Bungay, L.: In: Advances in Applied Microbiology (Eds. W. W. Umbreit and D. Perlman) p. 269, Academic Press (1968).
Canale, R. P.: Biotechnol. Bioeng. **11**, 887 (1969).
Canale, R. P.: Biotechnol. Bioeng. **12**, 353 (1970).
Constantinides, A., Spencer, J. L., Gaden, E. L.: Biotechnol. Bioeng. **12**, 803 (1970).
Corrans, I. J., Harris, B., Ralph, B. J.: J. South African Inst. Min. Met. **72** (8), 221 (1972).
Dean, A. C. R., Rogers, P. L.: Biochim. Biophys. Acta **148**, 280 (1967).
Dean, A. C. R., Moss, D. A.: Biochem. Pharmacol. **20**, 1 (1971).
Distefano, G. P.: "Hybrid Computation in the Process Industries", EAI, New Jersey (1968).
Fan, L. T., Shah, P. S., Erickson, L. E.: Simulation of Biological Processes by Analog and Digital Computers, Kansas State University Report No. **20**, 118 pp (1970).
Goodwin, B. C.: Temporal Organization in Cells, Academic Press (1963).
Goodwin, B. C.: Advan. Enzyme Regulation **3**, 425 (1965).
Gray, P. P., Dunhill, P., Lilly, M. D.: J. Ferm. Tech. (Japan) **50** (6), 381 (1972).
Gray, P. P., Rogers, P. L.: Biochim. Biophys. Acta **230**, 393 (1971).
Gray, P. P., Rogers, P. L.: Biochim. Biophys. Acta **230**, 401 (1971).
Handbook of Analog Computation, Electronics Associates Inc., West Long Branch, New Jersey (1971).
Hanson, T. P., Tsao, G. T.: Biotechnol. Bioeng. **14**, 233 (1972).
Harrison, D. E. F.: J. Cell Biol. **45**, 4574 (1970).
Harrison, D. E. F., Topiwala, H. H.: In: Advances in Biochemical Engineering **3** (Eds. T. K. Ghose, A. Fiechter and N. Blakebrough) Springer-Verlag, p. 167 (1974).
Higgins, J. J.: In: Computers in Biomedical Research, V. 2, Academic Press, N. Y. (1965).
Hockenhull, D. J. D., McKenzie, R. M.: Chem. Ind. **607** (1968).
Knorre, N. A.: Biochem. Biophys. Res. Commun. **31**, 812 (1968).
Koga, S., Humphrey, A. E.: Biotechnol. Bioeng. **9**, 375 (1967).
Korn, G. A., Korn, T. M.: In: Electronic Analog and Hybrid Computers, McGraw Hill, N. Y. (1972).
Lacey, D. T., Lawson, F.: Biotechnol. Bioeng. **12**, 29, (1970).
Levine, L.: In: Methods for Solving Engineering Problems using an Analog Computer, McGraw Hill, N. Y. (1964).
Luedeking, R., Piret, E. L.: J. Biochem. Microbiol. Technol. Eng. **1**, 393 (1959).
Milsum, J. H.: Biological Control Systems Analysis, McGraw Hill (1966).
Nyiri, L. K.: In: Advances in Biochemical Engineering **2** (Eds. T. K. Ghose, A. Fiechter, N. Blakebrough), Springer-Verlag p. 49 (1972).
Pring, M.: In: Concepts and Models of Biomathematics. Simulation Techniques and Methods. (Ed. F. Heinmets), M. Dekker Inc., p. 88 (1969).
Pye, E. K.: Can. J. Botany **47**, 271 (1969).
Reed, G., Peppler, H. J.: In: Yeast Technology, AVI Publ. (1973).
Röpke, H., Riemann, J.: Analog Computer in Chemie and Biologie, Springer-Verlag (1969).
Rogers, P. L.: Simulation of Microbial Systems with EAI–180 Analog/Parallel Logic Computer, EAI Applications (1972).
Rogers, P. L., Hendy, N.: A kinetic model for microbial leaching of nickel sulphide (NiS) In preparation (1976).
Rogers, P. L., McDonald, I. J.: Kinetic analysis of the growth of Streptococcus cremoris. In preparation (1976).
Rosenbrock, H. H., Storey, C.: Computational Techniques for Chemical Engineers, Permagon Press (1966).
Schultz, J. S., Gerhardt, P.: In: Bacteriol. Rev. **33**, 1–47 (1969).
Selinger, B.: Simulation of Chemical Systems with the EAI-180Analog/Parallel Logic Computer, Electronic Associates Pty. Ltd. (1973).
Shu, P.: In: Fermentation Technology Today (Ed. G. Terui), Fourth Int. Ferm. Symp. 183 (1972).
Snyder, M. F.: Trans. Bio-Med. Eng. **16** (4), 325 (1969).
Terui, G.: J. Pure Appl. Chem. **36**, 377 (1973).

Tuovinen, O. H., Kelly, D. P.: Z. Allg. Mikrobiol. **12** (4), 311 (1972).
Unden, A.: Computer controlled optimization of growth parameters in continuous culture, LKB Fermentation Symposium, Moscow (1974).
Unden, A., Heden, G. C.: In: First European Conference on Computer Process Control in Fermentation. INRA, Dijon (1973).
Williams, T. J.: Ind. Eng. Chem. **62** (12), 94 (1970).
Wong, C., Scharer, J. M., Reilly, P. M.: Can. J. Chem. Eng. **52** (5), 645, (1974).
Yano, T., Koga, S.: Biotechnol. Bioeng. **11**, 139 (1969).
Young, T. B., Bruley, D. F., Bungay, H. R.: Biotechnol. Bioeng. **12**, 747 (1970).
Young, T. B., Bungay, H. R.: Biotechnol. Bioeng. **15**, 377 (1973).
Zines, D. O., Rogers, P. L.: Biotechnol. Bioeng. **12**, 561 (1970).

CHAPTER 4

Preparation and Properties of Gel Entrapped Enzymes

K. F. O'DRISCOLL,
Dept. of Chemical Engineering, University of Waterloo, Waterloo, Ontario/Canada

With 6 Figures

Contents

1. Introduction

An enzyme can be immobilized by fixing it on the surface of a macroscopic material or by trapping it inside a matrix which is permeable to the enzyme's substrate. A valuable bibliography [1a] on all types of immobilized enzymes lists mor than 400 references from 1972 and 1973 alone. Abviously, the subject is of rapidly expanding interest, and it has been comprehensively treated by Zaborsky [1b]. For reasons discussed below, immobilization by trapping seems to have gained an unnecessarily poor reputation compared to immobilization by binding to a surface. It is the purpose of this review to survey the several entrapping procedures and to consider the qualities of the resulting entrapped enzyme with respect to their utility in various applications.

Any consideration of application for an immobilized enzyme system must be made in the light of an understanding of the following attributes of the system:

1. Enzyme loading
2. Kinetic behavior
3. Stability
4. Reactor configuration.

Although these attributes are highly coupled, they can be discussed in a general sense as separate factors. What follows is written primarily with entrapped enzymes in mind, but some of it is necessarily applicable to other types of immobilized enzymes.

Enzyme loading refers not only to the actual amount of enzyme which is entrapped, but it also includes the fact that in an entrapment procedure some enzyme may be lost due to leakage or denaturation. As a consequence of this, the immobilized preparation may have a specific activity per unit weight which is less than might be expected from the amount of enzyme used in the preparation. Furthermore, the amount of enzyme which can be entrapped in some cases is seriously limited by the solubility of the enzyme in the entrapping reaction mixture.

Other factors which may determine the actual specific activity of the immobilized preparation include unequal partitioning of the substrate between the immobilized phase and the supernatant liquid, slow diffusion of the substrate to the active site, microenvironment effects such as localized pH or ionic strength inside the entrapping matrix, and diffusive or partitioning effects on inhibitors of the enzyme activity. All these factors are here lumped under the term kinetic behavior.

The stability of an immobilized enzyme in use or storage may be one of the most important parameters in the cost estimation of an immobilized enzyme process. An immobilized enzyme can denature while immobilized, but it may do so at a rate higher or lower than the native enzyme. In the case of enzymes immobilized by entrapment the worry also exists that the enzyme may continuously leak out of the matrix, thus reducing the specific activity of the preparation and possibly contaminating the product.

Finally, consideration must also be given to the reactor configuration which is to use the immobilized enzyme. This factor is highly coupled to the others, in that the specific activity, kinetic behavior and stability are all quantitative factors which must be employed in optimizing a reactor configuration. In addition, the mechanical characteristics of the immobilizing material will determine its suitability for use in a given reactor: a soft, compressible material might cause too large a pressure drop across a tubular reactor and plug it; a hard, friable material might not withstand the shear field in a well-stirred batch or continuous tank reactor; a hydrated membrane in which an enzyme is entrapped may need to be supported. From a more positive viewpoint; an entrapped enzyme is not subject to bacterial attack; the mechanical strength of the immobilizing material can be controlled by, for example, preparing a polymer gel having a desired degree of hydration; multiple enzyme systems can easily be co-entrapped, or even whole cells may be entrapped.

From the above discussion it is apparent that careful attention must be given to a number of inter-related factors if an entrapped enzyme system is to be used, or considered for use, in a real process. Similar considerations, of course, must be made if a surface-immobilized enzyme is to be used. A choice of a particular immobilization technique, whether for a surface-immobilized enzyme or an entrapped enzyme, should be made on the basis of knowledge of all these factors. No single type of immobilization procedure can be expected to be perfect, and therefore the choice of an immobilization procedure ought to depend on a careful balancing of its known advantages and disadvantages.

In the recent past, almost all entrapped enzymes were immobilized in poly(acrylamide) gels. These gels are quite weak in a mechanical sense and have an open network that often permits the enzyme to leak out. As a consequence, entrapped enzymes have a poor reputation. This reputation is undeserved since enzymes may be entrapped in a variety of ways, some of which can lead to preparations that are most desirable in that their properties are superior for the particular use for which the immobilized enyzme is intended.

There are many, chemically different hydrogels which can be prepared with enzymes entrapped in them. Because the substrate must diffuse through the gel before it can be acted upon by the enzyme, a brief discussion of the diffusive processes in a gel is in order.

Yasuda *et al.* [2] have given an excellent description of how molecules diffuse in a hydrogel. If, in the preparation of the gel, a set of pores, permanent in space and time, be created, then molecules smaller than the minimum diameter of the pores can diffuse through them. If the diffusing molecule is bigger than the pores, then diffusion can still take place but it requires cooperative motion of the polymer chains to bring an element of free volume (void space) adjacent to the diffusing molecule. Under the driving force of a concentration gradient, diffusing molecules have net movement in one direction, and free volume elements move in the others. This has been termed activated diffusion.

Figure 1 shows schematically how the extent of hydration and the diffusing species molecular weight affects the net rate of diffusion.

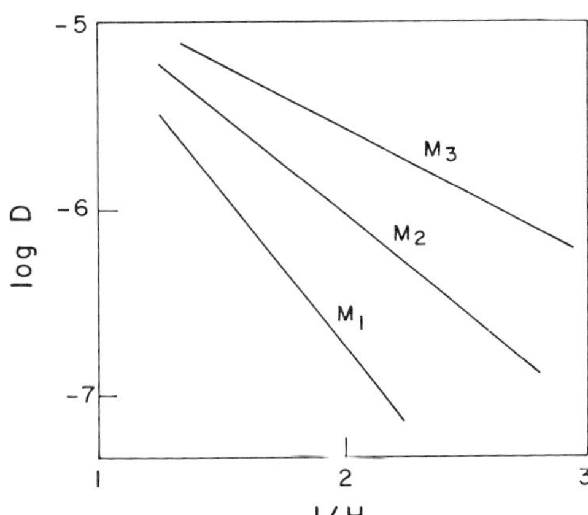

Fig. 1. Influence of extent of hydration (H) and molecular weight of permeant (M) on diffusivity (D) in a hydrogel; $M_1 > M_2 > M_3$

2. Preparation of Entrapped Enzymes

2.1.1 Gel-Entrapment by Polymerization

To entrap an enzyme one must either form a crosslinked polymeric network around the enzyme molecule, or place an enzyme inside a polymeric material and then crosslink the polymer chains. Bernfeld and Wan [3] appear to have been the first to use the former technique when they dissolved one of several enzymes (trypsin, ribonuclease, β-amylase, etc.) in an aqueous solution of N,N'-methylenebisacrylamide (BIS) and carried out a low temperature, free radical polymerization of this cross-linking monomer (Fig. 2). The choice of this particular monomer is easy to understand, since the

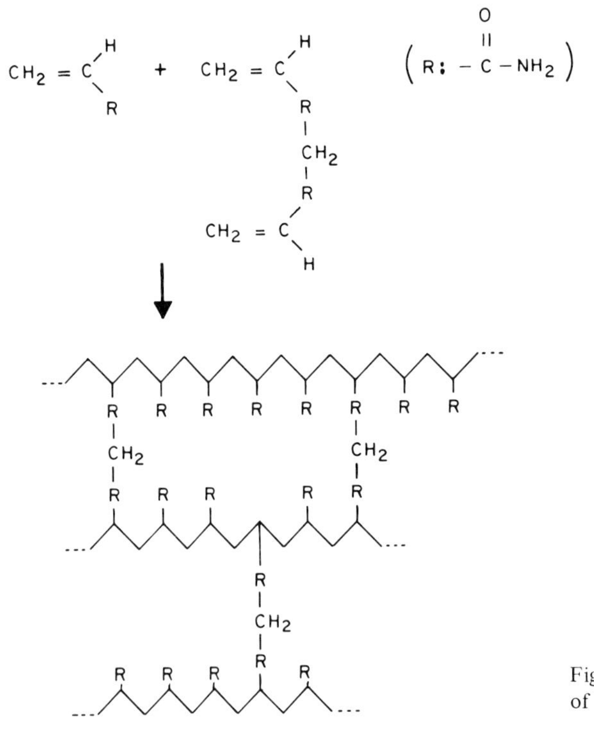

Fig. 2. Crosslinking copolymerization of BIS and acrylamide

preparation of hydrated gels for electrophoresis using acrylamide and BIS is a common biochemical laboratory procedure. Subsequent workers followed this lead and used these two monomers exclusively. In retrospect, it is unfortunate that other monomers were not used, since poly(acrylamide) gels crosslinked with BIS have extremely high (ca. 80–95%) water content and are therefore structurally weak and easily leak some enzymes. Nevertheless, much has been learned about gel-entrapped enzymes using poly(acrylamide) gels, and this knowledge can be carried over to preparations using other, more desirable monomer/polymer systems. So far, the only other vinyl monomer which has been investigated for entrapping enzymes is 2-hydroxyethyl methacrylate (HEMA) crosslinked with ethyleneglycol dimethacrylate (EGDMA) [4–6]. This is surprising when the large number of water-soluble, easily polymerized vinyl monomers is considered. Also surprising is the fact that no attempt appears to have been made to entrap enzymes by carrying out organic condensation polymerizations in aqueous media.

2.1.2 Solution Polymerization

The ingredients common to all gel-entrapping polymerizations which have been reported are: monomer, crosslinking agent, initiator, enzyme and aqueous buffer. If a polyfunctional monomer such as BIS or EGDMA is used, the monomer and crosslinker are the same. Some preparations have included a substrate for the enzyme or other material (such as albumin) added to protect the enzyme.

Polymerizations are commonly carried out in the absence of oxygen and at low temperatures, 0 to 25° C, in order to guard against thermal denaturation. Vinyl polymerizations are exothermic, and so the polymerization must be carried out slowly or be well thermostatted to minimize temperature rise. At the conclusion of the polymerization, which may take from 30 min to several hours, the hydrated gel contains the entrapped enzyme and, possibly, non-trapped enzyme, residual monomer and unused initiator. Common practice is to cut the gel into small pieces and wash with a buffer. Depending on the intended use, the gel may then be used as is, or dried, ground into small particles, sieved into known size ranges, and stored prior to rehydration and use.

Any or all of the components of the polymerization mixture can serve to denature the enzyme while the reaction is going on. Therefore their concentrations and relative proportions should be carefully chosen. Some general principles can be discerned from reported work, but each system is different.

One might suspect, *a priori*, that the free radicals which initiate the polymerization might also attack and denature the enzyme. Fortunately, this does not appear to be the case, probably because the concentration of free radicals in a vinyl polymerization seldom goes above 10^{-6}M. The substance of greatest danger to the enzyme activity is the monomer itself. Figure 3 shows the effect on activity of the enzyme which can be

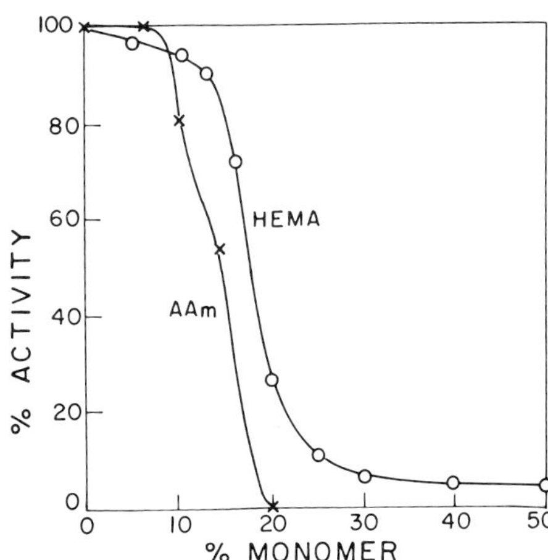

Fig. 3. Cholinesterase activity after 30 min at 25° C in acrylamide [7] (×). Trypsin activity after 12 hrs at 22° C in HEMA [8] (○)

observed by incubating the enzyme with a monomer solution [7, 8]. After only 20 min of incubation in a 20% solution of acrylamide (AAm) at 25° C cholinesterase was found to lose all its activity; similarly, trypsin incubated in aqueous HEMA solutions for 12 hrs showed great loss of activity at concentrations greater than 20%. Such information suggests that monomer concentration should be kept low in a polymerization recipe. However, as Degani and Miron showed [7], the amount of protein entrapped will fall off monotonically as the monomer concentration is decreased in the entrapping reaction mixture. The specific activity of the enzyme may pass through a maximum because of these two countering influences of monomer concentration as shown in Fig. 4.

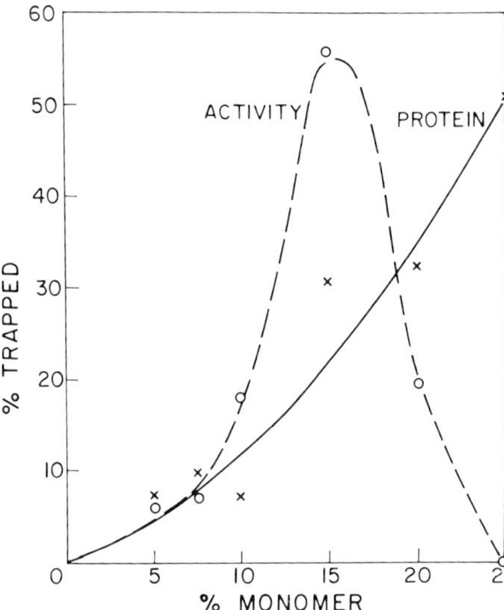

Fig. 4. Influence of monomer concentration on amount and activity of cholinesterase entrapped in a poly(acrylamide) gel [7]

In a similar vein, it appears that the crosslinking agent concentration may also be optimized. Figure 5 shows the data for the effect of BIS in a poly(AAm) gel on cholinesterase activity [7] and for EGDMA in a poly (HEMA) gel on glucose oxidase activity. [9] The choice of initiator and its optimal concentration does not lend itself to easy generalization. Low temperature redox systems, such as potassium persulfate and tetramethylene diamine (TEMED), are useful for low temperature work, and do not appear to injure enzymes in the low concentrations necessary for initiation of polymerization. The amount of enzyme dissolved in the polymerization reaction mixture is usually limited only by its solubility. Since the entrapping mixture typically contains 20 to 50% monomer, the enzyme solubility is not necessarily as high as it would be in pure buffer. This limitation on the maximum loading of enzyme in the final gel may prove to be the most serious factor limiting the use of gel-entrapped enzymes in some commercial processes. However, in those cases where a high loading is not necessary or where only a small amount of enzyme is available, gel entrapment provides a facile means for immobilizing enzymes with a minimum of preparative effort.

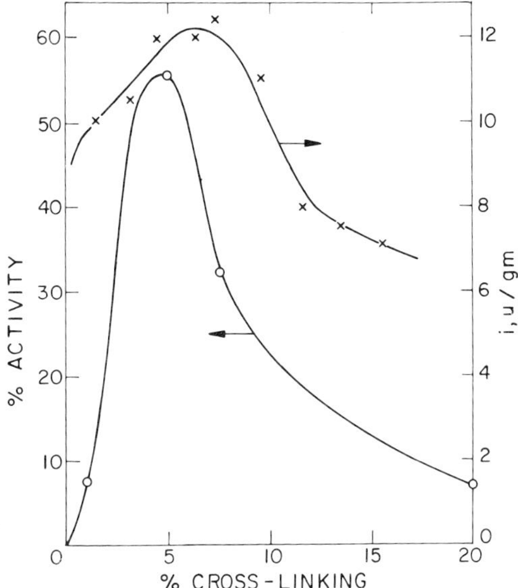

Fig. 5. Effect of crosslinking on activity of cholinesterase in poly(acrylamide) [7] (○), and glucose oxidase in poly(HEMA) [9] (X)

2.1.3 Suspension Polymerization

There have been a number of reports on gel entrapment by polymerization where the procedure varied sufficiently from the above to be worthy of note. For example, instead of carrying out an aqueous solution polymerization, the hydrogel entrapping the enzyme has been formed in a suspension polymerization [10, 11]. In such a polymerization, an aqueous phase containing monomer, initiator and enzyme is dispersed in a continuous, hydrophobic liquid. The resulting hydrogel consists of spherical particles, whose size can be controlled, to some extent, by the polymerization reaction conditions. The spherical particles have a potential advantage over the irregular, broken particles which can be obtained from a solution polymerization, in that the spheres pack more netly in a tubular reactor.

An interesting post-entrapment treatment was performed by Beck and Rase [11] on their suspension polymerized, gel-entrapped glucoamylase. They incorporated amylopectin, a very high molecular weight polymer, in their gel by dissolving it in the reaction mixture. It was therefore entrapped along with the enzyme. After the spheres were formed, they treated them with soluble glucoamylase. It is presumed that the soluble glucoamylase digested the amylopectin and thus developed a number of pores leading from the surface of the spheres toward its interior. These pores make the entrapped glucoamylase more accessible to a diffusing, low molecular weight substrate; the authors did note an increase in the activity of their gel to a partially hydrolysed cornstarch.

One difficulty with suspension polymerization is that a detergent is often used to stabilize the suspension. The use of such a detergent may denature some enzymes, so care must be taken in adopting existing recipes for suspension polymerization.

2.1.4 Whole-Cell Immobilization

Several successful immobilizations of whole-cell cultures have been reported where acrylamide [12–14] or HEMA [15] was used as the entrapping medium. Mosbach [12] pointed out that this provided a means of entrapping small amounts of enzymes or enzymes whose existence in the cell was transient. Slowinski and Charm [14] trapped a suspension of *Corynebacterium glutamicum* in poly(AAm) and noted that its activity continued for more than 3 months in producing glutamic acid. They also noted the problem of providing sufficient oxygen to the gel when it was used in a column. The whole-cell immobilization technique may also require the supplying of enzyme cofactors to replenish those low molecular weight species that are eluted from the gel.

2.1.5 Radiation-Initiated Polymerization

A concern with the possible effect of chemical polymerization initiators on the enzyme has led some workers to use irradiation as a generator of free radicals for the polymerization. Among the irradiation sources are included X-rays [16] and γ-rays [17]. In the case of the γ-irradiation, Maeda and co-workers [17] did not use any crosslinking monomer and yet formed a crosslinked gel of poly(AAm) which retained 30% of the activity of the invertase they were entrapping. Dobo [16], on the other hand, noted that high dosage of X-rays decreased the immobilized enzyme activity. It would appear that irradiation might be a technique of choice for initiating polymerization in certain cases, especially where a chemical initiating system is not possible (e.g. at very low temperature) or desirable because it would denature the enzyme. The amount of irradiation necessary to initiate polymerization can probably be expected to be less than that which would seriously inactivate the enzyme.

2.1.6 Inorganic Gels

Almost all of the immobilization by entrapment has been done in organic hydrogels, but Dickey [18] may have been the first to prepare an entrapped enzyme. In a study of the influence of molecules present during the formation of silica gels on their adsorbent properties, he prepared some gels with urease and with catalase that retained some activity. More recently, Johnson and Whatley [19] prepared a hydrogel by polymerizing a silicic acid sol in the presence of trypsin. They noted that the gel had 34% of the trypsin activity toward a low molecular weight amino-acid ester, but would not hydrolize a very large substrate, casein. They suggested [19] that inorganic gels might provide a better entrapment medium than organic ones on the basis of their resistance to microbial attack and that swelling and porosity would not be a function of pH. In actuality, organic gels are not particularly susceptible to microbial attack (very few polymers are), and the swelling and "porosity" can be optimized for a given, desired pH by addition of comonomers.

2.2.1 Gel Entrapment by Crosslinking Polymers

If an enzyme is dispersed in a linear polymer or polymer solution and the polymer is then crosslinked to form an insoluble network, the enzyme can be expected to be en-

trapped. At least four different types of preparation of this sort have been reported.
Poly(dimethylsiloxane) was crosslinked with tin octoate by Brown and co-workers [20] forming silicone membranes to immobilize several different enzymes. The success of the immobilization is difficult to establish from the reported data, since only relative activities are given. Guilbault and Das, however, have reported [21] that the immobilization conditions were too rigorous for cholinesterase since 80% activity was lost in their silicone preparations. Considering the hydrophobic character of this rubber, it is most likely that in these preparations low activity is a result of restricted permeability of the membrane to the water soluble substrate.

Guilbault and his co-workers have also immobilized a variety of enzymes in starch [21, 22]. They prepared the gel by dissolving the enzyme in a warm starch solution and then expressed the water from the cooled starch gel in a polyurethane pad that was used in analytical work.

Maeda and co-workers, who used radiation initiated polymerization [17] as described above, have also used γ-irradiation and electron beam techniques to crosslink an aqueous poly(vinyl alcohol) solution containing glucoamylase or invertase [23]. In the case of γ-irradiation no leakage of protein from the gel was observed, but there was not much enzyme activity either. Electron beam irradiation seemed more promising, since they recovered 50% of the enzyme activity and still had no enzyme leakage. The gels formed in this fashion are crosslinked because the high energy irradiation creates free radicals on the polymer chain which couple with each other. The amount of radiation required is somewhat greater than that needed for initiation of polymerization, and there is, therefore, more likelihood of damage ot the enzyme.

Although a considerable body of knowledge exists on the formation of gels by mixing polyvalent ions with polyelectrolytes, there has been only a single report [24] of using this technique to entrap enzymes. This technique, termed coarctation, would appear to be one of the more gentle ways of crosslinking polymers, and ought to be given more attention.

2.2.2 Reticulation

It must be remenbered that enzymes are polymers also, and the crosslinking of an enzyme directly will provide an insoluble polymer having enzymatic activity. This process, termed reticulation, has been well reviewed in Chapter 4 of Zaborsky's book [1b] and is mentioned here for the sake of completeness. In some cases, reticulated enzymes have been prepared while the enzyme was in a gel [25], or prior to the formation of a gel by polymerization [26]. In either case, the result is to create a "superpolymer" out of the enzyme, and thus reduce its mobility in the gel to the point where leakage does not occur. As Vieth has shown [25], this technique permits one to put an extremely high amount of enzyme into a membrane.

2.2.3 Miscellaneous Techniques

Microencapsulation has been pioneered by Chang [27] as a method for immobilizing enzymes. In his systems, a solution of enzyme is surrounded by a very thin-skinned sphere of polymer. Preparation of these microcapsules involves some art and substrate must diffuse through the polymer, but the spheres offer a number of advantages in use such as their high surface area, low pressure drop for a bed of them, and potential for high loading. Another form of microencapsulation has been reported [27c] where lipid spherules have been used to trap droplets of enzyme solutions. Such a technique might prove generally useful for oil-soluble enzymes.

An interpenetrating network (IPN) has been used in one unique investigation [28]. An esterase was isolated from a potato extract using electrophoresis on a poly(AAm) gel. The section of gel containing the enzyme was then perfused with more AAm monomer and that was polymerized. The resulting IPN held the enzyme quite tightly giving a good yield and high stability. This technique holds promise for immobilizing enzymes available only in small quantity.

Enzymes have also been entrapped within the pores of wet-spun synthetic fibres [29]. An aqueous enzyme solution is added to a polymer solution which is immiscible with water, and the extrusion of the resulting emulsion into a coagulating bath results in the fibre-entrapped enzyme. In principle, any polymer which can be wet-spun can be used to entrap one or more enzymes; in practice the choice of system strongly influences enzyme activity and stability.

3. Properties of Gel-Entrapped Enzymes

3.1 Kinetic Behaviour

There are several attributes of a gel-entrapped enzyme system which will affect the kinetics of the system. If we begin by considering a substrate molecule in solution outside a gel in which the enzyme is contained, we must sequentially recognize the possibility of a boundary layer diffusive barrier, of unequal partitioning of the substrate between gel and supernatent liquid, and of an internal diffusive barrier. Further, there may be microenvironmental effects inside the gel which may make the electrostatic field and/or the pH quite different at the active site form that which would be encountered in homogeneous solution of the native enzyme. If there is product or substrate inhibition or activation of the enzyme, some, or all, of these same effects may also be important on that account.

Kobayashi and Laidler [30] have given a most comprehensive theoretical treatment for *solid* supported enzymes with diffusive and electrostatic effects taken into account for reaction on the surface. They point out that their treatment will also hold for reaction inside a gel-entrapped enzyme assuming there is no internal mass transfer resistance. Unfortunately such an assumption will seldom be warranted. As discussed below, there are numerous treatments of the internal diffusion effect on gel-entrapped enzyme kinetics, but little has been done on the *combined* problem of internal and external diffusion resistance for immobilized enzymes. The very simple reaction, $A \rightleftarrows B$, has been treated [31a] for catalysis inside a homogeneous particle; the derivation gives insight into the complexity of the problem and the form which a solution might follow for gel entrapped enzymes. A numerical solution considering external mass transfer limitations as well as internal diffusion has been presented by Hamilton *et al.* [31b]. In a reactor configuration where the external effects can be eliminated by high flow or stirring rates, the problem can become unimportant.

The influence of diffusion within the gel on enzyme kinetics has been considered both experimentally and theoretically. It is generally recognized that no analytical solution exists for treating Michaelis-Menten kinetics in a gel where diffusion exists, except for the limiting cases of substrate concentration $[S]$ being very much higher or lower than the value of the Michaelis constant K_m' inside the gel. This value is usually similar in magnitude to K_m for the native enzyme in solution. The result [32] for the limiting condition of high substrate concentration $[S]$ in a membrane is that the rate in the gel v' is proportional to the concentration of enzyme in the gel:

$$v' = k_c'[E]_{gel} \tag{1}$$

where k_c' is the catalytic constant of the enzyme in the gel. [1]

When $[S] \ll K_m'$, the rate equation is far more complex:

$$v' = \frac{k_c'[E]_{gel}[S]}{K_m'} F \tag{2}$$

where the function F is given for a membrane as

$$F = \frac{2}{\alpha l} \frac{\cosh \alpha l - 1}{\sinh \alpha l} \tag{3}$$

The quantity α is a constant for a given set or reaction conditions and relates the reaction rate to the diffusion constant D:

$$\alpha = \left(\frac{k_c'[E]_{gel}}{K_m'D}\right) \tag{4}$$

for spherical particles of radius R, a very similar solution gives [33]

$$F = \frac{3}{(\alpha R)^2} (\alpha R \coth(\alpha R) - 1) \tag{5}$$

where R is the sphere's radius.

It should be noted that $(\alpha l)^2$ or $(\alpha R)^2$ is the Damköhler number, β, and is related to a Thiele modulus ϕ by

$$\phi = (\beta K_m'/([S] + K_m'))^{1/2} \tag{6}$$

Thus, it is obvious that the kinetic treatments of gel-entrapped enzymes can lean heavily on existing ways of regarding heterogeneous catalysis.

Numerical solutions for those cases where the substrate concentration is not very much greater or less than K_m' have been presented [31b, 34–37]. Moo-Young and Kobayashi [34] give a complete theoretical treatment in terms of an effectiveness factor which is defined as the ratio of the rate with the immobilized enzyme to the rate that would be obtained at the same substrate concentration and with no internal mass transfer limitation.

They treat not only simple Michaelis-Menton kinetics but also consider product and substrate inhibition, and show that the effectiveness factor will generally be less than unity for moderate substrate concentrations. If substrate inhibition occurs, the effectiveness factor may exceed unity and become quite large.

O'Driscoll and Korus [37] presented a similar treatment for spherical gel particles and showed experimental data which conform to the predictions of the theory. Lineweaver-Burke plots were noted to be curved, as theory predicts, and render suspect the estimation of K_m' values from such plots. Figure 6 shows the variation of the effectiveness factor with $[S]$ at various values of the Damköhler number β.

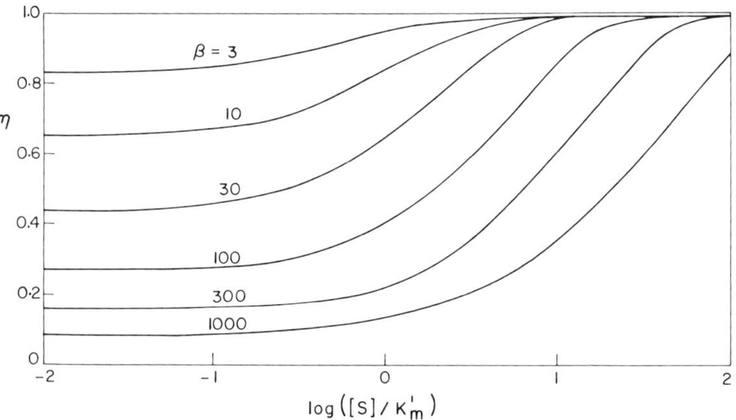

Fig. 6. Variation of effectiveness factor, η, for enzymes (entrapped in spheres) with dimensionless substrate concentration $[S]/K_m'$ and Damköhler number, β [8]

From these works and a number of others it is now clear that diffusion in a gel membrane or sphere affects the immobilized enzyme reaction kinetics in a manner which may be quantitatively treated in terms of a Thiele modulus or Damköhler number. The effects of extra-gel diffusion can be minimized by high flow rates and the effects of internal diffusion can be quantitatively considered in the classical terms of heterogeneous catalysis. The simple fact remains that gel-entrapped enzymes are expected to be useless for substrates with low diffusivities. This would exclude those substrates of molecular weight appreciably greater than 500 or 1 000.

Partitioning of substrate between the supernatant and the gel has been explicitly included in most of the theoretical treatments, but little experimental information exists to suggest that it is very important in the absence of electrostatic effects. In the author's laboratory partition coefficients have usually been found to very from unity by only 20 or 30% for uncharged substrates.

The influence of electrostatic effects in a gel on its enzyme activity has been extensively treated by Goldman *et al.* [38]. Polymeric gels can easily be made with ionic functional groups on the main chain by incorporating, for example, acrylic acid or vinyl pyridine as a comonomer. Therefore, the existence of ionic effects offers a great opportunity for modifying the activity of an enzyme. Similarly, the gel may be made to provide a microenvironment which is internally buffered so that the active site experiences a pH different from that of the supernatant. This could then shift the pH optimum of the enzyme. Although there are numerous experimental reports of such effects being observed, they have usually been adventitious rather than the result of a deliberate attempt to modify the microenvironment. There are exceptions to this, e.g., the inclusion of acrylic acid as a comonomer with acrylamide by Bech and Rase [11] to buffer their gel-entrapped glucoamylase.

The measurement of the catalytic and the Michaelis constants has often been done without regard to diffusive effects, therefore they are best labelled as apparent constants, $k'_{c(\mathrm{app})}$ and $K'_{m\,(\mathrm{app})}$. Even so, it appears that these constants have been only slightly affected by immobilization in gels. Often $K'_{m\,(\mathrm{app})}$ values are reported as somewhat greater than the K_m value for the native enzyme, and the result attributed to internal diffusion effects. In a study of glucose oxidase immobilized in a poly(HEMA) gel O'Driscoll and coworkers [39] showed that the rate constants for decomposition of the glucose-enzyme complex and for the oxidation of the reduced form of the enzyme were the same as those for the same enzyme immobilized on glass or in free solution, but that the apparent bimolecular rate constant for interaction of glucose and the enzyme was intermediate in value between the glass-immobilized and free solution values. It was suggested that this might be attributed to a change in the microenvironment induced by the chemical nature of the gel.

Undoubtedly, as more quantitative work is reported in which diffusive effects are rigorously eliminated, the picture will become clearer, and we shall have a quantitative idea of how real and important microenvironment effects are.

3.2 Stability

The thermal or chemical denaturation of an immobilized enzyme may occur at a rate higher or lower than that of the native enzyme in solution because it is in a different microenvironment. The polymer gel surrounding the active site may also be expected to protect the enzyme from bacterial attack. Therefore it is reasonable to expect that a gel-entrapped enzyme might last longer in use than an enzyme in solution. Data which have been reported in the literature usually relate only to thermal denaturation, and it is not obvious from them that gel entrapment will always lower the fundamental rate of thermal denaturation. In some studies, at high temperatures, the gel-entrapped enzyme has been found to denature faster than the free enzyme in solution. Other work has shown little or no increase in stability as a result of immobilization. A qualification of this statement must be made however, since some use applications may be carried out at temperatures low enough that the rate of thermal denaturation is too low to study. In such a situation, the use of an accelerated aging test, achieved by raising the temperature, may not be valid.

An additional complication is the influence of intraparticle diffusion effects on the *apparent* denaturation rate of a gel-entrapped enzyme, which has been pointed out by Ollis [40]. Because diffusion reduces the apparent activity in a manner which changes non-linearly with entrapped enzyme concentration in the gel, a decrease of the number of active sites due to denaturation will not result in a proportionate decrease in the gel's activity. Therefore, the apparent half-life for denaturation will be greater for the gel than for the native enzyme by a factor of as much as two. This effect does not seem to have been considered in any published experimental data.

3.3 Mechanical Properties

There has, as yet, been little research directed at examining the mechanical properties of gel-entrapped enzymes. However, since these properties will be almost exclusively those of the gel, it is reasonable to extrapolate from our knowledge of gels. Because they are crosslinked and plasticized by the water of hydration, gels are elastic and easily deformed. They fail easily in high shear, tension or compression. Therefore it is unreasonable to expect a gel-entrapped enzyme to be usable in large columns with high pressure drops or in highly stirred reactors. They will be useable as particles in small columns, in membranes, possibly supported to provide strength, or in a "pellicular" form as suggested by Horvath [41]. A pellicular catalyst particle would consist of an interior core of high mechanical strength surrounded by a coating of the gel entrapped enzyme. Such a material would not be expected to withstand abrasion, but it could be used in large packed columns.

The influence of attrition of fine particles on the kinetics of enzyme covalently bound inside cellulose particles has recently been discussed by Lilly *et al.* [42], but not much else been written on the mechanical aspects of immobilized enzyme preparations.

4. Multiple-Enzyme Systems

If gel-entrapped enzymes have one outstanding attribute, relative to other immobilization techniques, it is that they come closer to mimicking the natural enzyme in the cell. Since multiple enzyme systems are so common in cells, it is probable that complex syntheses will be better carried out by immobilized enzymes which are trapped in gels.

Published work on multienzyme immobilizations is just beginning to appear in quantity, and most of it comes from Broun, Thomas and coworkers [43] and from Mosbach and his colleagues [44]. A recent paper of Mosbach *et al.* [44] will serve as an example: a three-enzyme system was used as a model for oxalacetate production and utilization in mitochondria. When NADH was reoxidized in this system (by the addition of pyruvate in the slurry reactor) the rate of the coupled, 3-enzyme system immobilized in Poly(AAm) was four-fold faster than the free enzymes in solution, and twice as fast as the same system covalently bonded onto Sephadex. Care was taken to ensure equivalent single-enzyme activities and that partitioning effects were absent.

Simply put, rate enhancement is understood to occur in gel-entrapped, multienzyme systems because the intermediates, products of one enzyme and substrates for the next, are at an abnormally high local concentration relative to what would be the case in free solution. A quantitative appreciation of this phenomenon can be gained from the theoretical work by Lawrence and Okay [45]. They examined the equations describing the system of glucose oxidase and catalase co-immobilized and homogeneously distributed throughout a porous support. The glucose would be oxidized to produce H_2O_2 and the catalase regenerate oxygen. It was shown that, at a low Thiele modulus and a non-zero ratio of peroxide to oxygen, there would be an expected rate enhancement for glucose oxidation of the order of two-to five-fold.

Some very early workers on gel-entrapped enzymes also had an appreciation of the relationship between naturally occurring enzymes and the gel systems. As early as 1964, van Duijn *et al.* studied enzymes immobilized in poly(AAm) films as models for cytochemical determination of enzymes [46]. They appear to have been the first to appreciate the diffusion control of gel-entrapped enzyme kinetics and attempted to utilize it in their work.

Given the refinement of immobilization techniques, so well demonstrated by Mosbach's group [44], it is probable that the use of multiple enzyme systems in gels will advance rapidly.

5. Qualitative Considerations

The applications of immobilized enzymes may be conveniently divided into three categories: process catalysis, analytical determinations and clinical or biomedical uses. Gel-entrapped enzymes have potential advantages in some of these applications, and some very specific disadvantages or limitations in others.

For process catalysis, gel-entrapped enzymes suffer from their mechanical weakness as described above. In addition, the enzyme loading which can be achieved in some preparations is too low to make them attractive for large volume use. However, in processes where the volume of production is small, and the dollar value added is high (such as specialty pharmaceuticals), the ease of immobilization by gel entrapment or the ability to entrap whole cells or multienzyme systems may make it the procedure of choice. The speculative tone of the foregoing is made necessary by the generally slow acceptance of immobilized enzyme technology of all kinds.

Analytical devices employing gel-entrapped enzymes have been the subject of a number of reports, and some will undoubtedly be commercially available in the very near future. Updike and Hicks described a blood-sugar analysis using immobilized glucose oxidase and a polarographic oxygen electrode [47]. The use of gel-entrapped enzymes in conjunction with specific ion electrodes has recently been reviewed [48]. Trace analysis of pesticides or metals is possible using the inhibition or activation of enzyme activity [49]. In all such applications, the stability of the preparation and its specificity is more important than the enzyme loading. In addition, special qualities of the gel can be provided to make the immobilized preparation most suitable for the use. For example, poly(HEMA) gels are known to be quite compatible with blood, and gel-entrapped glucose oxidase has been used in the author's laboratory, in a manner similar to Updike and Hicks, for continuous analysis of whole blood.

Possible clinical uses of enzymes which have been immobilized range from administration of the enzyme to a patient as a therapeutic agent to cure disease, to attempts to replace enzymes which are missing because of genetic defects [50]. The most notable therapeutic use is of asparaginase, which has been shown to be effective in inducing remission in patients with acute lymphocytic leukemia. Unfortunately, the enzyme is a foreign protein and causes an antibody reaction. Chang has used microencapsulated asparaginase in mice, where the capsules were injected intraperitoneally [51]. This technique is more efficient than an extracorporal device, and the enzyme does not induce an immune response since the microcapsule is impermeable to the high molecular weight antibodies. Ohnuma *et al.* have reported [52] data similar to Chang's using asparaginase immobilized in a poly(HEMA) gel. Updike *et al.* have used asparaginase in a poly(AAm) gel and have given a preliminary report on its physiological fate in rats [53]. The use of immobilized enzymes, gel-entrapped or other kinds, in enzyme-deficient diseases appears at present to be limited by the problem of getting the immobilized preparation to the site where it is needed.

From the above it is obvious that the use applications of gel-entrapped enzymes are few at present, in spite of the tremendous research efforts of the last few years. The "futuristic" [54] character of this technology is still there, waiting to be put to use. Regardless of the exact use it is important that the immobilized enzyme be regarded as an integral part of a *system* containing not only the enzyme and the substrate but also the immobilizing phase and all the rest of the environment in which it is being used. This emphasis on an immobilized enzyme as part of a system is essential because those who work in this field are necessarily crossing disciplinary boundaries, and it is all too easy to ignore the importance of areas other than one's own.

Nomenclature

D = diffusion coefficient of substrate in gel
$[E]_{gel}$ = active enzyme concentration per unit volume of gel
F = function defined by Eq. (3)
H = fractional extent of hydration of gel
k'_c = catalytic constant of enzyme in gel
K'_m = Michaelis constant of enzyme in gel
l = membrane thickness
M = molecular weight of substrate
R = radius of spherical gel particles
$[S]$ = substrate concentration in gel
v' = rate of enzyme catalyzed reaction per unit volume of gel
α = defined by Eq. (4)
β = Damköhler number = $\alpha^2 l^2$, or = $\alpha^2 R^2$
η = effectiveness factor, i. e.: rate catalyzed by gel entrapped enzyme relative to rate of free enzyme under equivalent conditions
ϕ = Thiele modulus (Eq. 6)

References

1a. "Immobilized Enzymes: A Compendium of References ..." Corning Glass Works, Corning, N. Y., 1972.
1b. Zaborsky, O. R.: "Immobilized Enzymes" CRC Press, Cleveland O., 1973.
2a. Yasuda, H., Peterlin, A., Colton, C. K., Smith, K. A., Merrill, E. W.: Makromol. Chem. **126**, 177 (1969).
2b. Yasuda, H., Lamaze, C. E., Peterling, A.: J. Polymer Sci. **A-2, 9** 1117 (1971).
3. Bernfeld, P., Wan, J.: Science **142**, 678, (1963).
4. O'Driscoll, K. F., Izu, M., Korus, R.: Biotechnol. Bioeng. **14**, 847, (1972).
5. Miyamura, M., Suzuki, S.: Nippon Kagaku Kaishi, 7, 1274, (1972).
6a. Turkova, J., Hubalkova, O., Krivakova, M., Coupek, J.: Biochim. Biophys. Acta. **322**, 1, (1973).
6b. Gould, F. E., Ronel, S. H.: Germ. Pat. **No. 2, 305**, 320 (1973).
7. Degani, Y., Miron, T.: Biochim. Biophys. Acta. **212**, 362, (1970).
8. Korus, R., Ph. D. Thesis: University of Waterloo, Waterloo, Ontario, (1974).
9. O'Driscoll, K. F., Kapoulas, A.: unpublished work.
10. Nilsson, H., Mosbach, R., Mosbach, K.: Biochim. Biophys. Acta **268**, 253, (1972).
11. Beck, S. R., Rase, H. F.: Ind. Eng. Chem. Prod. Res. Develop. **12**, 260, (1973).
12. Mosbach, K., Mosbach, R.: Acta Chem. Scand., **20**, 2807, (1966).
13. Mosbach, K., Larsson, P.-O.: Biotechnol. Bioeng. **12**, 19, (1970).
14. Slowinski, W., Charm, S. E.: Biotechnol. and Bioeng. **15**, 973, (1973).
15. Turkova, J., Hubalkova, O., Krivakova, M., Coupek, J.: Biochim. Biophys. Acta. **322**, 1, (1973).
16. Dobo, J.: Acta. Chim. Acad. Sci. Hung. **63**, 453, (1970) through CA **72**: 96979 m (1970).
17. Maeda, H., Yamauchi, A., Suzuki, H.: Biochim. Biophys. Acta. **315**, 18, (1973).
18. Dickey, F. H.: J. Phys. Chem., **59**, 695, (1955).
19. Johnson, P., Whateley, T. L.: J. Colloid Interface Sci. **37**, 557, (1971).
20a. Brown, H. D., Patel, A. B., Chattopadhyay, S. K.: J. Biomed. Mater. Res. **2**, 231, (1968).
20b. Pennington, S. N., Brown, H. D., Patel, A. B., Knowles, C. O.: Biochim. Biophys. Acta. **167**, 479, (1968).
20c. Pennington, S. N., Brown, H. D., Patel, A. B., Chattopadhyay, S. K.: J. Biomed. Mater. Res. **2**, 443, (1968).
21. Guilbault, G. G., Das, J.: Anal. Biochem. **33**, 341, (1970).
22a. Bauman, E. K., Goodson, L. H., Guilbault, G. G., Kramer, D. N.: Anal. Chem. **37**, 1378 (1965).
22b. Guilbault, G. G., Kramer, D. N.: Anal. Chem., **37**, 1675, (1965).
23. Maeda, H., Suzuki, H., Yamauchi, A.: Biotechnol. Bioeng. **15**, 607, 827, (1973).
24. Horvath, C., Sovak, M.: Biochim. Biophys. Acta. **298**, 850, (1973).
25. Wang, S. S., Vieth, W. R.: Biotechnol. Bioeng. **15**, 93, (1973).
26. Brown, G., Thomas, D., Domurado, D., Berjonneau, A., Guillon, G.: Biotechnol. Bioeng. **15**, 359, (1973).
27a. Chang, T. M. S.: Science **146**, 524 (1964).
27b. Chang, T. M. S.: "Artificial Cells", C. C. Thomas Publ., Springfield Ill., (1972).
27c. Gregoriadis, G., Leathwood, P. D., Ryman, B. E.: FEBS Letters **14**, 95, (1971).
28. Dinish, K. N., Shivaram, K. N., Stegeman, H.: Z. Anal. Chem. **263**, 27, (1973).
29. Dinelli, D.: Process Biochemistry **7(8)**, 13, (1972).
30. Kobayashi, T., Laidler, K. J.: Biotechnol. Bioeng. **16**, 77, (1974).
31a. Smith, N. L., Amundson, N. R.: Ind. and Eng. Chem. **43**, 2156, (1951).
31b. Hamilton, B. K., Gardner, C. R., Colton, C. K.: A. E. Ch. E. J. **20**, 503, (1974).
32a. Goldman, R., Kedem, O., Katchalski, E.: Biochem. 7, 4518, (1968).
32b. Sundaram, P. V., Tweedale, A., Laidler, K. J.: Can. J. Chem. **48**, 1498, (1970).
33. O'Driscoll, K. F., Hinberg, I., Korus, R., Kapoulas, A.: J. Polymer Sci. **C 46**, 227 (1974).
34. Moo-Young, M., Kobayashi, T.: Can. J. Chem. Eng. **50**, 162, (1972).

35. Fink, D. J., Na., T. Y., Schultz, J. S.: Biotechnol Bioeng. **15**, 879, (1973).
36. Engasser, J. M., Horvath, C.: J. Theor. Biol. **42**, 137, (1973).
37. O'Driscoll, K. F., Korus, R.: Can. J. Chem. Eng. **52**, N 775 (1974).
38. Goldman, R., Goldstein, L., Katchalski, E.: "Biochemical Aspects of Reaction on Solid Supports", G. R. Stark, Ed., Chap. 1, Acad. Press, N. Y. (1971).
39. Hinberg, I., Kapoulas, A., Korus, R., O'Driscoll, K.: Biotechnol. Bioeng. **16**, 159, (1974).
40. Ollis, D. F.: Biotechnol. Bioeng. **14**, 871, (1972).
41. Horvath, C., Engasser, J. M.: Ind. Eng. Chem. Fundam. **12**, 229, (1973).
42. Regan, D. L., Dunnill, P., Lilly, M. D.: Biotechnol. Bioeng. **16**, 333 (1974).
43a. Brown, H. D., Patel, A. B., Chattopadhyay, S. K.: J. Chromatogr., **35**, 103, (1968).
43b. Broun, G., Thomas, D., Selegny, E.: J. Membrane Biol. 8, 313, (1972).
44a. Mosbach, K., Mattiasson, B.: Acta Chem. Scand. **24**, 2093, (1970).
44b. Brown, H. D., Patel, A. B., Chattopadhyay, S. K.: J. Chromatogr. **35**, 103, (1968).
44c. Mattiason, B., Mosbach, K.: Biochim. Biophys. Acta **235**, 253, (1971).
44d. Gestrelius, S., Mattiasson, B., Mosbach, K.: Biochim. Biophys. Acta **276**, 339, (1972).
44e. Gestrelius, S., Mattiasson, B., Mosbach, K.: Euro. J. Biochem **36**, 89, (1973).
44f. Srere, P. A., Mattiasson, B., Mosbach, K.: Proc. Nat. Acad. Sci. U.S.A. **70**, 2534, (1973).
45. Lawrence, R. L., Okay, V.: Biotechn. Bioeng. **15**, 217, (1973).
46a. von der Ploeg, M., van Duijn, P.: J. Roy. Microscop. Soc. **83**, 415, (1964).
46b. von Duijn, P., Pascoe, E., von der Ploeg, M.: J. Histochem. Cytochem. **15**, 631, (1967).
47. Hicks, G. P., Updike, S. J.: Anal. Chem. **38**, 762, (1966), "also U. S. Patent 3, 788, 950 (1974)."
48. Gough, D. A., Amdrake, J. D.: Science **180**, 380, (1973).
49. Towshend, A.: Process Biochemistry, **8(3)**, 23, (1973).
50. "Enzyme Therapy in Genetic Diseases", D. Bergsma Ed., National Foundation – March of Dimes, White Plains, N. Y., (1973).
51. Chang, T. M. S.: Enzyme **14**, 95, (1972/73).
52. Ohnuma, T., O'Driscoll, K., Korus, R., Wolczak, I.: J. Pharm. Exper. Therap. **195**, 382 (1975).
53. Updike, S., Prieve, C., Magnuson, J.: Enzyme **14**, 77, (1972/73).
54. Rubin, D. H.: "A Technology Assessment Methodology: Enzymes (Industrial)", National Technical Information Service, Springfield, Va., (1971).

Professor *K. F. O'Driscoll*
Dept. of Chemical Engineering
University of Waterloo
Waterloo, Ontario/Canada

Related Titles

Advances in Biochemical Engineering

Editors: T. K. Ghose, A. Fiechter, N. Blakebrough

Volume 1: With 70 figures. V, 194 pages. 1971

Volume 2: With 70 figures. III, 215 pages. 1972

Volume 3: With 119 figures. IV, 290 pages. 1974

There is a pressing need to explore the area of interaction between engineering principles and biological systems—a research field which until recently has been neglected. The intricacies of biotechnology embrace the full range of present-day chemical engineering and the biological sciences. Their interaction has not only brought about a great reduction in the cost of producing some foods and life-saving drugs, but has advanced new concepts to improve the understanding of the engineering principles involved in the operation of processes like air sterilization, separation of biological molecules, enzyme systems, continuous culture, handling of biological fluids, oxygen transfer in large-scale fermentation systems, etc. The problem is to find a realistic solution bridging these two distinct areas; this requires the breaking down of communication barriers between biological scientists and engineers and the active interchange of information. The series Advances in Biochemical Engineering is designed toward this end.

Related Titles

Progress in Molecular and Subcellular Biology

Volume 1: 32 figures, VII, 237 pages, 1969.
By B. W. Agranoff, J. Davies, F. E. Hahn, H. G. Mandel, N. S. Scott, R. M. Smillie, C. R. Woese.
This volume contains contributions by Woese, Davies and Mandel who are concerned either theoretically or experimentally with the nature of the genetic code and its correct translation. Smillie and Scott write about the biosynthesis of chloroplasts and the photoregulation of the underlying basic processes. Agranoff reviews critically current ideas concerning the molecular basis of higher nervous activities. The managing editor, Hahn, makes a critical analysis of the origin, conceptual content and probable future developments in molecular biology.

Volume 2: 158 figures, IX, 400 pages, 1971.
Proceedings of the Research Symposium on Complexes of Biologically Active Substances with Nucleic Acids and Their Modes of Action. Held at the Walter Reed Army Institute of Research, Washington, 16–19 March 1970.
These 28 contributions are a definite account of the most recent research results on selected aspects of the title topic. The major topical subdivisions are: Antibiotics and Nucleic Acids–Antimalarials and Nucleic Acids–Alkaloids, Natural Polyamines and DNA–Intercalation into Supercoiled DNA–Synthetic Drugs and Dyes Binding to Nucleic Acids–Antimutagens–Carcinogens and Nucleic Acids–The Natural State of DNA. Among the contributors are D. M. Crothers, G. F. Gause, W. and H. Kersten, L. S. Lerman, H. R. Mahler, P. O. P. Ts'o, J. Vinograd and M. Waring. The editor, F. E. Hahn, introduces Volume 2 with a review of the history, current state and perspectives of the field of nucleic acid complexes. The book will interest research scientists active in the study of nucleic acid complexes and molecular biologists, molecular pharmacologists, biochemists and biophysicists seeking comprehensive and up-to-date information in this specialized field.

Volume 3: 58 figures, VII, 251 pages, 1973.
By A. S. Bravermann, D. J. Brenner, B. P. Doctor, A. B. Edmundson, K. R. Ely, M. J. Fournier, F. E. Hahn, A. Kaji, C. A. Paoletti, G. Riou, M. Schiffer, M. K. Wood.
The volume is concerned with the transcription and translation of genetic information and with products of both processes in prokaryotic and eukaryotic organisms. Reverse transcription is reviewed with respect to its theoretical importance to molecular biology and its practical importance to cancer research. Translation, i.e. protein biosynthesis, is analyzed in the light of knowledge of numerous inhibitors of individual reaction steps in the overall sequence. The transcription of transfer RNA is reviewed in detail as one aspect of transcription of genetic information. Thalassemia and Bence-Jones proteins have been treated as selected examples of molecular pathology in eukaryotes. The nature of mitochondrial DNA in neoplastic cells is reviewed as one contribution to the molecular biology of cancer. The book signifies the advancement of molecular biology from the study of bacteria and their viruses to the consideration of events and entities in eukaryotic organisms.

Springer-Verlag Berlin · Heidelberg · New York